This intriguing series encompasses exciting trends and discoveries in areas of human exploration and progress: astronomy, anthropology, biology, physics, geology, medicine, health, genetics, and evolution. Sometimes controversial, these timely volumes present stimulating new points of view about our universe . . . and ourselves. Among the titles:

Ancestral Voices: Language and the Evolution of Human Consciousness
Curtis G. Smith

The Human Body and Why It Works
Raymond L. Powis

Toward a New Brain: Evolution and the Human Mind
Stuart Litvak and A. Wayne Senzee

On Civilized Stars: The Search for Intelligent Life in Outer Space
Joseph F. Baugher

THE BOOK OF THE MOON

THE BOOK

THOMAS A. HOCKEY

OF THE MOON

A Lunar Introduction to Astronomy, Geology, Space Physics, and Space Travel

PRENTICE HALL PRESS • New York

Published by Prentice Hall Press
A Division of Simon & Schuster, Inc.
Gulf + Western Building
One Gulf + Western Plaza
New York, NY 10023

Library of Congress Cataloging-in-Publication Data

Hockey, Thomas A.
The book of the moon.

(Frontiers of science)
Bibliography
Includes index.
1. Moon. I. Title. II. Series.
QB581.H63 1986 523.3 86-472
ISBN 0-13-079963-7

Manufactured in the United States of America

10 9 8 7 6 5 4 3 2 1

First Edition

To Ruby Hockey,
who started me on this project
many years ago

ACKNOWLEDGMENTS

I wish to acknowledge Professor Reta Beebe for the suggestion that led to the course upon which this book is based as well as for her support and advice during every phase of its production.

I would like to express my appreciation to Elizabeth Gillette for enabling me to teach *A Return to the Moon* in 1983. I also want to thank the Baskett family: Keath, Murray, and the Kaypro 2 microcomputer.

Much of the responsibility for turning my manuscript into something intelligible went to Andrea Dobson-Hockey, without whom this work would not have been possible.

CONTENTS

PREFACE

This book addresses the two faces of the Moon: astronomical wonder and strange, new world. I have written it for anyone who has ever stood and watched the Moon rise.

The first chapter discusses the Moon in our imagination. Subsequent chapters describe the evolution of our understanding of the Moon from our observation post here on the Earth. Along the way, I outline those particular concepts and inventions that have brought the Moon closer to us. An example is the discussion of the telescope in Chapter 4.

Later chapters of this book look at the exploration of space from a historical perspective: how we got to the Moon, what we did once we got there, and what we learned. As for what we learned, many of the results from the Apollo experience were not immediately known or adequately translated into public understanding. Few laymen today would be able to satisfactorily answer the question, "What did we learn about the Moon by going there?" This question is addressed at length, especially with new information that has been gleaned from the Apollo and Soviet data only in the 1980s.

This book is not an in-depth account. Rather, it is a brief, nontechnical review of humanity's affair with the Moon for those who have an "armchair" curiosity about science and about history. Numbers and calculations are kept to a minimum. Key words are in bold face and are included in the glossary at the back of the book.

This review is appropriate now, not only because we can begin to see lunar science in its historical perspective, but because this may be the "calm before the storm" in lunar studies. Only recently have scientists and others begun to propose a new beginning for lunar exploration, a beginning that will be made possible by the technology explosion of the 1970s and 1980s. This time, the emphasis will be on what we can do with the Moon. The answers are many. This, then, is a summary to date, a look at what has been done, before we return to the Moon.

1 INTRODUCTION

On the north side of a small arroyo draining into the Rio Grande, I have found an Indian petroglyph chipped into the soft patina covering the local rocks. The shape is that of a crescent. Next to it, another petroglyph depicts a rabbit.

The Mogollon Indians, who inhabitated this area between 200 B.C. and A.D. 1200, frequently made images of the animals they hunted. The Moon was often a symbol of fertility to native American peoples. Could it be that a pre-Columbian artist was illustrating a prayer for more plentiful prey in the next hunt?

Interpretation of ancient rock art is speculative at best. Still, it is possible that the crescent figure on that rock face indeed represents the Moon. The crescent appears rarely in nature. Nowhere is it more obvious that at night when it calls our attention to the changing shape of the Moon. The crescent is the universal symbol for the Moon. It has appeared whenever people have sought to record their thoughts about the Moon on stone or on paper, up to and including the present.

Veneration of the Moon is also universal. Its appearance in mythology spans every continent from the jungles of Central America to the Ionian Peninsula, from the Fertile Crescent of Asia to Australia. It is not surprising that the Moon became an object of worship. It is spectacularly placed in the night sky. It was different from everything else the ancients knew about, and its true nature was a complete mystery. Today, there are few places on Earth where the Moon is regarded as a deity. Today, the Moon is a place. Through the study of astronomy, it has become a world not totally unlike our own. Since our ventures into space in the 1960s, the Moon has become a place to visit.

Has this new reality of the Moon affected our perception of it? Sadly, these days fewer and fewer persons take time really to look at the Moon. As life on Earth becomes more complex, many fear that they must keep a steady eye on it lest it change beneath their feet.

The Moon is too familiar. There is a tendency to take it for granted since the hype and fervor surrounding Project Apollo and this country's goal of landing a man on the Moon. This goal became an end in itself, and once it was achieved, people were ready to go onto something else. Indeed, our voyages to the Moon ended, at least for a while, not long after they began.

It has now been longer since men last set foot on the Moon than the entire length of time from the first man in space to Apollo XI. A student beginning college today is too young to remember the excitement when the world turned on its television sets to watch Neil Armstrong step onto the lunar surface. Indeed, there are children in high school today who had not been born by July 20, 1969.

These students look on the race to the Moon as history, and films of men in their primitive spacecraft amuse them as quaint and anachronistic. Now we are in the age of the Space Shuttle, the most sophisticated vehicle ever created. Computer technology that once navigated spacecraft from the Earth to the Moon now controls games and toys. Such marvels are expected; our imaginations ask for something more. Children no longer play with astronaut dolls and put together scale models of American spacecraft. Luke Skywalker and the fantasy of "A long time ago, in a galaxy far, far away . . ." have (temporarily?) replaced the real conquest of space.

Yet there is still something special about the Moon. Young people will still, given a chance, gaze at the Moon in wonder. The pale, alluring luminosity of the Full Moon is still inspiring. Perhaps it is even more so now that its reality as a neighbor in space to travel to and explore has become apparent. The Earth's satellite and the journey to it may still capture our imagination more than the technological feats being performed routinely today in Earth orbit.

This may be so because exploration has always excited human beings: from the first tentative steps in the direction of the next oasis to the great voyages of the Renaissance. More recently, there has been the search for the sources of great rivers and the conquest of remote mountaintops and the Earth's poles. Only the depths of the oceans and space remain uncharted. The near-Earth-orbiting Space Shuttle is a truck, albeit an impressive one, not a vessel of discovery. The Moon is the tangible frontier beyond our world.

2
THE MOON IN OUR IMAGINATION

MOON WORDS

The word *Moon* comes from the Anglo-Saxon *Mona* and is a cousin of the German *Mond.* The Greek word *selene* has become our prefix for the study of the physical Moon **(selenography).** It also appears in the names of the element selenium, the mineral selenite, and selenicereus (Moon cactus), a plant native to tropical America.

The Latin term *Luna* is used as well. Lunar science encompasses aspects of selenography and selenophysics. One wonders what the ancient Romans would have made of expressions such as "lunar base" and "lunar module"!

About the first century B.C., the Sun, Moon, and five planets visible in the sky (Mercury, Venus, Mars, Jupiter, and Saturn) became associated with the days of the week. The Moon got the second one, right after the Sun. The word *Monday* comes from the Anglo-Saxon *Monan daeg,* which is a translation of the Latin *Dies Lunae* (Day of Luna). To the Germans, it is *Montag* (Day of the Moon); in French, it's *Lundi;* and in Spanish, it's *Lunes* (Luna's Day).

Colors were assigned to the planets as well. Gold was reserved for the Sun. The Moon's ghostly white light was emulated by the use of silver. Images of Moon gods were often adorned in silver. In as remote a place as Peru, an ancient temple contains silver plates with effigies of the Moon pounded into them. A superstition, which persists today, says that it is good luck to turn over silver coins in your pockets when you first see the New Moon.

Moonstone is a pearly mineral that the ancients made into amulets to promote health and fertility. Moonwart was a popular medicinal herb in the Europe of the Middle Ages.

Words like these remind us of how often the Moon's presence is found in our language even today. "Shoot the Moon," "The Cow Jumped Over the Moon," *Moon River, Moonlight Sonata,* mooning, moonbeams, moonshine, and Moon Zappa all come to mind. Lovers are said to be "moonstruck." We also still call people "loony."

MOON LORE

Early people were ignorant of the workings of most everything in their environment. That which they did not understand, they assumed to be under the direct command of some superior intelligence. This intelligence often took the form of a god to be worshipped.

The Moon, so different from Earthly, nonluminous bodies, was surely such a diety. This belief persisted throughout most of the history of civilization. The Moon was a favorite god of nomadic tribes who, traveling at night, looked to it for guidance and protection. As late as the fourteenth century, there were literate persons in Europe who believed that the Moon moved because it willed to do so. This long history of personification of the Moon has yielded a rich body of lunar mythology.

Lest we dismiss such beliefs too superciliously, it should be remembered that not all Moon lore is in the religious beliefs of antiquity or of primitive people in isolated corners of the globe. Astrology still abounds, and the worship of the Moon continued in Europe into the sixteenth century despite dire warnings of the consequences from the Christian church.

The Moon possesses obvious characteristics that were incorporated into its mythological personality. The

Moon's light is a cool one. It does not scorch the land as the Sun does. Therefore, it was considered beneficial to crops and livestock. The Moon was an agricultural god, and its worship often began before that of the Sun.

Because the Moon went though phases, it was sometimes considered to be a god of anything that was incomplete. Lunar deities were often pictured without a leg or hand by early artists. When the Moon god was male, he might appear castrated.

More frequently, though, the Moon was thought to be female and to have female qualities in contrast with the Sun, usually a male figure. Her particular qualities varied, however. The Moon goddess was sometimes a virgin and sometimes a prostitute! The Moon was a fertility goddess in keeping with its agricultural roots. Some Asian, Australian, and native American peoples believed that it was the Moon that could induce pregnancy in creatures, including humans. Lunar fertility rites were fostered by the knowledge that the human menstruation period roughly corresponds to a monthly cycle.

A third quality of lunar gods was a connection with the sea. It has been known since antiquity by coastal dwellers that the tides rise and fall with the Moon. Thus, the Moon god or Moon goddess has had aquatic associations. In Brazil, for instance, two themes were brought together in naming the Moon the "Mother of the Seas." An unlikely creature, the snail, was an early lunar symbol.

Some of the earliest lunar mythology has been passed down to us from the Sumerian civilization, which flourished along the Tigris and Euphrates rivers before the second millenium B.C.

In the Sumerian and later Babylonian pantheon, the Moon had several names—Nammu (or Nanna), Sin, and Hur—and was thought to be male. The holy place of Sin was the ancient city of Ur. Other sources put Sin's sanctuary in the west on Mt. Sinai, where presumably he rested after his long trip across the sky. Worship of Sin there may have continued into the Christian era.

Sin was the Lord of Wisdom and Ordainer of the Laws of Heaven and Earth. He was the keeper of time and the giver of dreams and oracles. The Moon was given precedence over the Sun in astronomical records that survive today. Sin may have preceded the Sun god in the religious hierarchy as well. A sacred day was dedicated to Sin: every seventh day. Sivanna (or Simanu), the third of the 12 Babylonian months, was dedicated to him.

Sometime between 2123 and 2100 B.C., the king Ur–Nammu began to build a temple at Ur to Sin–Nammu. This namesake king is responsible for one of the earliest marvels of ancient architecture. The three-step ziggurat at Ur was uncovered by archaeologists in the 1920s. Such structures were artificial mountains rising above the desert plains that were used for observing and being closer to the gods of the heavens.

One of the more interesting stories about this era was told by the Roman historian Josephus. He wrote that around 2000 B.C. the biblical Abraham visited the great ziggurat in his home town of Ur. (The temple was already more than 100 years old.) Abraham suggested to the priests there that if Sin were really an intelligent deity, he would move in a more conventional manner. He argued that because it did not, the Moon was under the authority of a higher God. The Sumerians were incensed by Abraham's theories, and the soon-to-be father of Israel decided to get out of the country for his own safety.

In another great ancient civilization of the Fertile Crescent—Egypt—the Moon goddess Nit was a secondary persona to the Sun god Ra. In fact, during the reign of King Akhenaton from 1379 to 1362 B.C., solar monotheism temporarily but entirely took away the Moon's deity status. This cult soon vanished, though, and Nit was returned to her place as goddess of spinners and inventor of the loom. Nit was a female character. However, it is interesting to note that pursuits such as the making of cloth were the province of men in Egypt. Nit traveled in a grand barge across the sky that was reminiscent of transportation down the lifeline of the Egyptians, the Nile.

Pigs became associated with the Moon in Egypt. In Lower Egypt, they were sacrificed at the time of the Full Moon. Another god, Osiris, had lunar aspects, although he was normally a grain deity. On the eve of the festival of Osiris, pigs were sacrificed and eaten. Poor people who did not have pigs made pig figures out of dough. During other times of the year, the eating of swine was considered taboo.

In the first millenium B.C., Semitic tribes inhabited the region that is now Israel, Syria, and Lebanon. Among these were the Israelites. The Syrian mother-goddess was Astarte, who had lunar associations. In Hebrew, she was Ashtoreth. The Syrians borrowed many of the traits of the earlier Sumerian Moon god. Astarte often appeared with the head of a cow. A crescent sat between its horns. Alternately, Astarte was a female figure holding a small pig.

Astarte was greatly revered. When a meteorite fell near a place in Syria, it was presumed to be Astarte herself. A temple was immediately erected on the spot.

The enthusiastic worship of Astarte caused trouble for the Israelites at the time of Moses. The Bible records numerous instances when their people were drawn to the wild rites of surrounding tribes. These rites were dedicated to Baal, the Sun god, or to Astarte. Recall the molten image of a calf of Old Testament fame. Later in the Bible, even King Solomon was seduced by a Sidonian woman who persuaded him to worship Astarte. He planted groves and erected altars to Astarte that were later destroyed by Josiah. In the first book of Samuel, when the Philistines came upon the body of Saul, they placed his weapons in the temple of Astarte as an offering. The cult of Astarte persisted and spread all over the Mediterranean. It even appeared in Britain.

The Greeks gave us the most enthralling lunar mythology of all. In their long history, not one but several gods were linked with the Moon. Even before the Greeks arrived, a cult of Artemis flourished in Asia Minor. Artemis was a lunar goddess of fertility and protector of the young. Her dual personality also made her the bringer of plagues and sudden death. The image of Artemis in her temple at Ephesus accentuates her female role by having several breasts. In her guise as Moon goddess, Artemis was clothed in a long, flowing robe. She wore a crescent diadem and carried a torch. By the time of Alexander the Great, the Temple of Artemis at Ephesus was considered one of the Seven Wonders of the World!

The three-headed goddess, Hecate, also had lunar qualities, particularily that of change. Her three faces represented those of the Moon. In fact, she often adorns illuminated lunar maps today. This symbolism also represented what the Greeks considered to be the three phases of womanhood. The New Moon was a young girl before puberty. The coming of the Full Moon represented her maturity. Its waning to crescent again depicted her post-child-bearing years and eventual death. Additionally, Hecate was a goddess of crossroads. Her three faces were the three choices a traveler must select from upon coming to crossroad.

Housewives would assemble on the night after the Full Moon to offer barley cakes to Hecate–Artemis. These cakes often held candles to replenish the moonlight. Moon cakes appear in various mythologies. The prophet Jeremiah rebuked his people for making cake offerings to the "Queen of Heaven." Moon cakes were also made in China and in the Pacific islands during festivals that occurred every eighth month. In the nineteenth century, citizens of Lancashire, England, still practiced the custom of baking barley cakes in honor of the Moon.

The Greeks differentiated between the goddess of the Moon and the Moon itself, Selene (or Mene). Selene did not figure greatly in Greek mythology. One existing tale has her falling in love with a handsome shepherd boy named Endymion. She later puts the boy into perpetual sleep, an allusion to another lunar aspect, the realm of slumber.

When the Romans borrowed Greek mythology wholesale, Selene became Luna. Later, however, Artemis, Hecate, and Selene were all combined into Diana, goddess of the moon, the hunt, and virginity.

In ancient China, the Moon was not deified. There, the Moon was respected as a fertility agent. To become pregnant in China was to become possessed by the Moon. In the fourth century, women of the imperial court conducted sexual exercises while exposing themselves to the Moon's light.

Elsewhere in the world, the Mayan civilization associated the Moon with sexual passion. The waxing Moon was a young, beautiful girl. As she became older, she became more malevolent. She now reigned over childbirth and was often pictured with snakes adorning her body. Today, Central American Indians describe the Moon as a pot repeatedly filled and then emptied.

Back in Europe, by the time of the late Roman Empire, astrology had largely replaced lunar cults. The influence of the heavenly bodies was now not thought to be directly the work of a deity, but these bodies were considered to be omnipotent all the same. Much of nature was controlled by the stars and planets. For instance, the Emperor Tiberius would allow his hair to be cut only when the Moon was waxing.

The astrological Moon was an emblem of magnetism and melancholy. It was the planet of dreamers and visionaries. It stood for phenomena of the night: hypnosis, imagination, hallucination, somnambulism, and madness.

The Moon was usually characterized as a cold, moist body. It governed the realm of things watery and phlegmatic. The Moon brought the morning dew. This view was elucidated by William Shakespeare in *A Midsummer Night's Dream,* Act III, Scene 1, where Titania says:

> The Moon methinks looks with a wat'ry eye,
> And when she weeps, weeps every little flower.

Because the Moon was somehow responsible for the rising and falling of the tides, its influence was extended over the flow of blood and other body fluids as well as the sap of plants. All through the Middle Ages, the Moon was supposed to have a significant effect on both agriculture and health. Various agricultural tasks, such as planting and harvesting, were believed to be most successful when done during the appropriate phase of the Moon.

The Moon was thought to influence the medicinal powers of certain plants. A healing herb might be effective only if it was picked in the appropriate manner on the right day of the month. When such an herb failed to cure a patient, an excuse could always be found by saying that the Moon had not been properly taken into account.

The Roman author, Pliny the Elder, recorded the old belief that the Moon has a narcotic property that causes persons exposed to its light long enough to become drowsy or stupified. Because the Moon's light is most obvious at night, it is not hard to see how this "effect" might have become associated with it. Pliny advocated using the Moon's rays as an anesthetic. He said that it was especially good to use when removing warts.

In the second century, the surgeon Galen taught that the Moon controlled the four bodily "humors." Children born when the Moon was full had high levels of these fluids and were destined to be robust and long-lived. Those born at that time of month when the Moon's influences were low would be sickly and die prematurely.

Galen's writings contain some curious "facts" about the Moon: the parts of the entrails of a mouse equal in number the days of the month; the centers of the eyes of certain farm animals contract and dilate with phases of the Moon!

The idea that things swell and shrink with the Moon was a common one. Not coincidentally, many of these were animals that live near the shore or in tidal pools. Fish, sea urchins, crabs, and tortoises were supposed to cyclically grow and then become lean inside their shells with the coming and going of the Moon. Pliny tells us to kill and eat tortoises on the fifteenth day of the month when they are fattest so as to get the most for our efforts. Even the Greek philosopher Aristotle (384–322 B.C.) considered the brain to be a cold, insensitive organ (a miniature room, really) that waxes and wanes sympathetically with the Moon.

The link between medicine and astrology held fast for more than a millenium. For instance, a prescription written by an apothecary in the Middle Ages might call for applying the liver of a cat that was prepared and killed under a waning Moon.

The German mystic Hildegard wrote that the absence of the Moon during its new phase or in the midst of an eclipse could cause epilepsy. Even the late-Renaissance thinker, Francis Bacon (1561–1626), reiterated astrological beliefs stated by Pliny that the brains of animals swell when the Moon is full.

The Moon also governed bleeding, one of the most popular medical procedures of the day. A fourteenth-century author warns other physicians not to bleed patients when the Moon is in the constellation of Gemini for fear that the tissue will not repair itself under these circumstances and the patient will bleed to death. Owing to the dubious benefits from bleeding, which often increased the risk of infection and shock, this advice might actually have been medically sound if it had been applied to every astrological sign!

The Moon's imagined influence extended even to life and death. A curious anecdote comes to us from the reign of Richard I in England. The author claims to have witnessed moonlight shining through a hole in a barn and focusing on the back of a horse, thereby somehow killing the horse's groom as he stood nearby!

Belief in the Moon's effects on the Earth persists today in secularized form. It is widely believed that the Moon influences the weather even though the tides in the atmosphere are negligible. Most people have heard of a "dry Moon" or a "wet Moon," terms that attempt to explain rain prediction based upon whether the horns of the Moon point upward to hold back the rain or downward to pour it out.

There are still gardeners who will tell you that potatoes should be planted during the New Moon because potatoes develop in underground darkness; and there are ranchers who shear sheep when the Moon is waxing so that the new wool will grow back rapidly. Other aphorisms include, "Hog's meat will not keep if it is left hanging where the Moon will shine on it" and "After a Moonlit Christmas, expect a lean harvest and a fat graveyard."

The Full Moon has long been a symbol of bizarre occurrences and is a stock feature in movies and books about werewolves, vampires, and other strange creatures of the night. The words *lunacy* and *lunatic* come from the belief, still popular, that mentally ill persons are more violent when the Moon is full.

3 THE MOON AND THE ANCIENTS

THE REAL MOON

In *Love's Labor's Lost,* Act IV, Scene 2, Shakespeare poses the ancient riddle:

> You are two bookmen. Can you tell by your wit,
> What was a month old at Cain's birth that's not five
> weeks old as yet?

If you cannot answer this riddle immediately, you can be assured that the solution will appear in the next few pages. First, though, we will take a look at the real, not mythical, Moon.

The Moon is 3,476 kilometers in diameter, seven-tenths the size of the planet Mercury. Its mean distance from the Earth is 384,401 kilometers; the actual distance varies

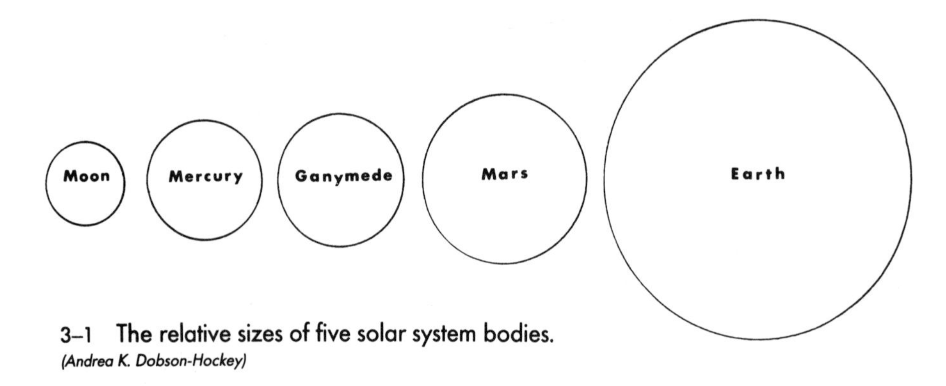

3–1 The relative sizes of five solar system bodies.
(Andrea K. Dobson-Hockey)

from 356,400 to 406,700 kilometers. (The metric system is used throughout this book for physical quantities. However, occasionally English units are used to give the American reader a better feel for the quantity or distance being described.) To give these numbers some meaning, imagine that the Earth is a softball (Figure 3–1). The Moon would then be a golf ball 12 feet away from it. It would take 9½ years to walk to the Moon, assuming certain difficulties involved with the trip were overcome! An automobile traveling on an imaginary road to the Moon at a constant 55 miles per hour would reach its destination in just under six months.

We call the Moon the Earth's satellite, but that is really a matter of perspective. The Earth and Moon orbit around a common center of gravity called the **barycenter** (i.e., the center of gravity of the Earth–Moon system). The barycenter is not at the center of the Earth but is a point 4,672 kilometers from the center of the Earth. The barycenter is 1,706 kilometers below us. The Moon is relatively massive compared to the moons of other planets, weighing in at 7.35×10^{22} kilograms. (Scientific notation is used in this book to denote very large and very small numbers. Recall that 10^{22} means the numeral 1 followed by 22 zeroes.) This compares with 5.98×10^{24} kilograms for the Earth. It may seem that the Earth is clearly the winner in its celestial tug-of-war with the Moon. It is certainly more massive, and the barycenter is closer to the center of the Earth than the center of the Moon. However, compared to other planet and satellite pairs in our solar system, ours is unique. Our Moon's mass is a larger fraction of the mass of its primary body than that of any other moon. The Moon's mass is 1.23% of the Earth's. By comparison, the mass of Triton, the planet Neptune's largest satellite, is 0.2% of that of Neptune itself. This is the nearest competing pair. (The mass of the planet Pluto's satellite, Charon, is not accurately known.) The Moon influences its parent body nearly 10 times more than a satellite of any other parent body does. Because of this near parity, the Earth and Moon are sometimes referred to as sister planets rather than planet and satellite (Figure 3–2).

(NASA photo)

3–2 The first photograph of the "Double Planet" was taken by the Voyager I spacecraft on September 16, 1977. The Earth is 12,000,000 kilometers away. The Moon is farther.

The Moon exhibits its dynamic phases because it orbits the Earth. At any given time, the Sun can illuminate only half of the lunar sphere. Thus, the geometry of the Earth, Sun, and Moon allows us to see various fractions of this illuminated portion. Phases are not caused by the Earth's shadow as some people believe.

At **New Moon,** the Sun, Moon, and Earth are almost in line (such an alignment is called a **syzygy**). The Moon is between the Sun and the Earth. The Sun shines on the backside of the Moon—that is, the side we cannot see. As the Moon moves eastward in the sky, we begin to see a thin crescent. As this crescent grows from night to night, the Moon is said to be *waxing* (from the Anglo–Saxon word *weaxon,* meaning "grow"). The Sun, Earth, and Moon form a right triangle 7.5 days after the New Moon. The bodies are now said to be in **quadrature.** We see half of the potentially visible disk of the Moon illuminated. This phase is called the **First Quarter.** As the lighted part of the disk appears to grow still larger, the Moon is said to be **Waxing Gibbous.** Finally, 15 days after New Moon, the Sun, Earth, and Moon are in syzygy again, but this time the Earth is in the middle and astronomers say that the Moon is in **opposition.** We see a fully illuminated disk, the **Full Moon.** The lighted part of the disk begins to recede as the Moon continues in its orbit. The Moon is now waning. It is **Waning Gibbous** until the quadrature point. This time, instead of seeing the west half of the disk, we see the east half. It is 22 days after New Moon, and the phase is **Third** (or **Last**) **Quarter.** The waning crescent is seen until the old Moon disappears in the morning glare of the Sun only to reappear in the evening several days later and repeat the cycle. (See Figure 3–3.)

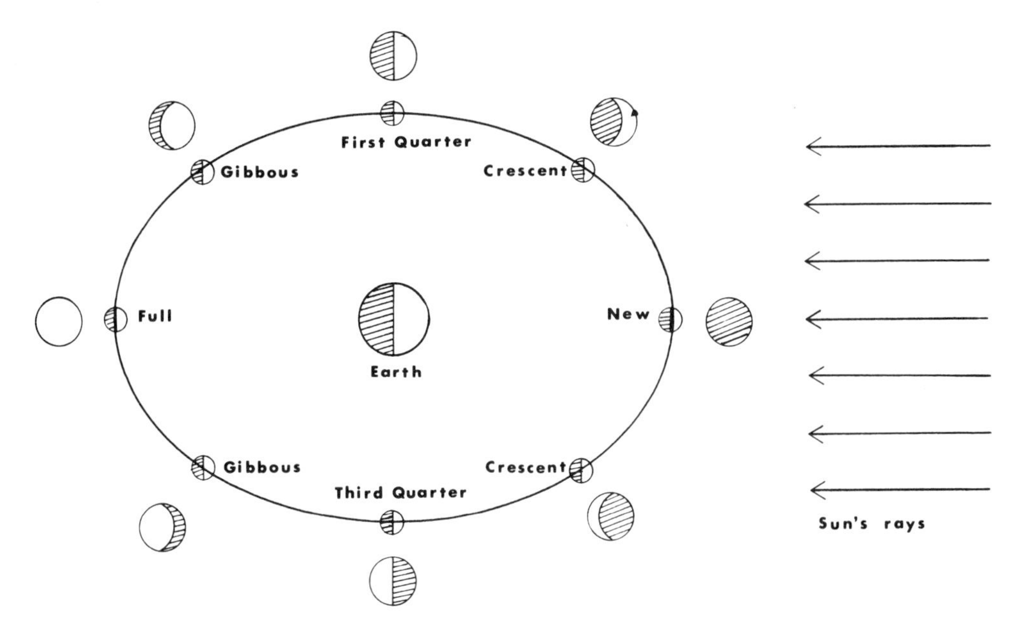

3–3 The phases of the Moon. The Sun always illuminates half of the lunar sphere. The position of the Moon in its orbit relative to the Earth-Moon line determines the phase that we see. The larger circles represent what the Moon looks like in the sky at different times of the month. *(Andrea K. Dobson-Hockey)*

The Moon orbits the Earth once every 27.32 days, when measured with respect to the seemingly fixed background of stars. This is the **sidereal** month. The word *month* comes from the Latin word *mensus* ("to measure"). After one sidereal month, the Moon will be seen among the same stars in the night sky.

The phases of the Moon are governed by the Moon's apparent position relative to the Sun. Since the Sun appears to move across the sky during the year in the same direction as does the Moon during the month, the Moon must catch up with the Sun from behind. Hence, it takes longer for the Moon to return to an apparent position relative to the Sun than to a fixed star. This month respective to the Sun's apparent motion is the **synodic month,** 29.53 days long.

The orbit of the Moon is inclined to the plane formed by the orbit of the Earth about the Sun (the **ecliptic**). The **inclination** of the Moon's orbit varies from 4 degrees, 59 minutes of arc to 5 degrees, 18 minutes of arc. Furthermore, it is not circular. The Moon's orbit is elliptical with the barycenter of the Earth–Moon system occupying one of the **focii** of the **ellipse.** (The **eccentricity** of this ellipse varies from 0.04 to 0.06.) Therefore, the Moon is closer to the Earth at one point in its orbit than at another. The point of closest approach is called the **perigee.** The farthest point is called the **apogee.**

The Sun's gravity affects the Earth–Moon system. One effect is to cause the line connecting the lunar perigee and apogee points to travel slowly eastward. This imaginary **Line of Apsides** precesses around the Moon's orbit once every 8.85 years. Thus, if one measures the month with respect to the moving Line of Apsides, yet a third result will be obtained. This month is called the **anomalistic** month and is 27.55 days long.

The points at which the Moon's orbit intersects the ecliptic are its orbital **nodes.** The node at which the Moon is traveling from south of the ecliptic to north is the **ascending node.** The other is the **descending node.** A line connecting these point is called the **Line of Nodes.** The Line of Nodes regresses westward with a period of 18.61 years. If one measures the length of the month relative to the Line of Nodes, one will get still another definition of the month—the **nodical** or **draconic month,** which lasts 27.21 days.

The Moon rises an average of 51 minutes later each night. However, around September, this **retardation** shrinks to 20 minutes. Hence, a Full Moon occurring during this time of year will be visible in the early evening for several nights in a row. The extra moonlight is appreciated by farmers who have given this particular Full Moon the name **Harvest Moon.** The Full Moon one month later is called the **Hunter's Moon.** The opposite effect occurs in March. This may be why we tend to associate the Full Moon with Fall leaves and Halloween pumpkins rather than with Spring flowers and Easter baskets.

The apparent diameter of the Moon is 32 minutes of arc, or 0.53 degrees. This means that the width (or height) of the Moon takes up 1/681 of a circle drawn all the way around the sky, about the same as that occupied by a penny held against the sky at a distance of seven feet. The apparent diameter of the Sun is nearly 32 minutes of arc, too! The Moon is much smaller than the Sun, but it is so much closer to us that the apparent sizes are the same. (*Warning:* Do not attempt to measure the size of the Sun by looking at a penny held up to it. Such a direct look at the Sun can damage your eyes.)

People tend to think of the Moon as being bigger than the Sun. This is partially because it is difficult to look directly at the Sun in order to make the comparison. Also, it is difficult to estimate the size of any object in the sky just by looking at it. Notice how artists often depict the Moon in paintings, as being unrealistically huge. The Moon seems to be larger when it is on the horizon than it does when high in the sky. Actually, the opposite is true, although the difference is not noticeable.

The wonderful coincidence in the apparent sizes of the Moon and Sun makes possible something that can be seen from no other planet in our solar system. This visible disk of our satellite will neatly fit over that of the Sun. When the Moon occults the Sun in this way, we have a solar **eclipse** (Figure 3–4). If the Moon passes directly over the Sun and blots out all of its light for a time, the eclipse is said to be **total.** If it sweeps by on either side so that only a fraction of the Sun is covered at any given time, the eclipse is **partial.**

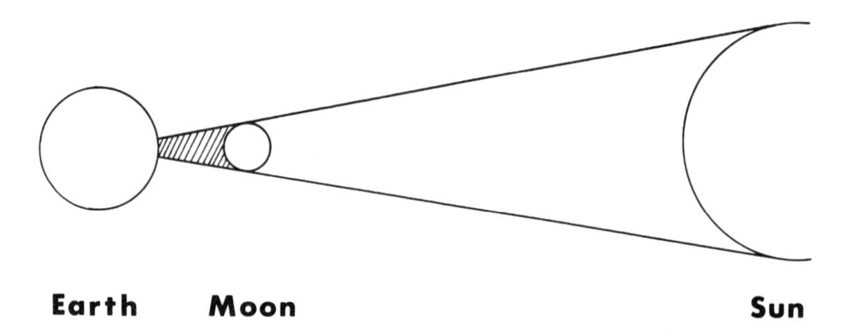

3–4 A solar eclipse. The hatched area represents the shadow of the Earth. In reality, the Moon would only cast its shadow over a small area on the planet.
(Andrea K. Dobson-Hockey)

Why do we not have a total solar eclipse at every New Moon? Recall that we said that at New Moon the Sun, Moon, and Earth are in line. They are not *exactly* in line, however. Remember that the orbit of the Moon is inclined to the ecliptic. The only time that all three bodies are in a straight line is when they are all in the ecliptic—when the Moon is at one of the nodes. The Moon is at a nodal point only twice on every orbit. For a solar eclipse to occur, the Moon must be exactly "new" to an observer on the Earth at the same time it is at one of its nodes. This coincidence does not occur frequently (Figure 3–5).

Furthermore, the shadow of the Moon on the Earth, that region where the Sun appears eclipsed, is very narrow. Its exact width is determined by the relative distance of the Moon from the Earth and Sun at the time of the eclipse. Both the Earth and Moon are sometimes closer to

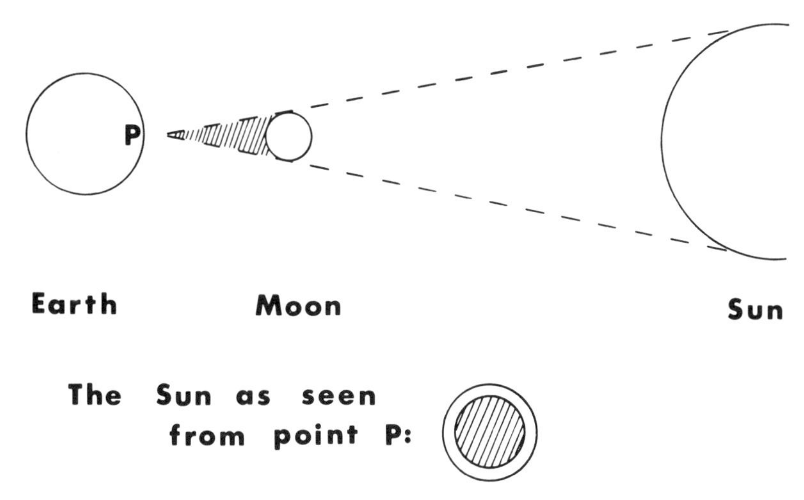

3–5 A total eclipse. The shadow cone of the Moon reaches the Earth. An observer watching the Sun from within the shadow will see a totally darkened disk, as depicted below.
(Andrea K. Dobson-Hockey)

the bodies around which they orbit (the Sun and Earth, respectively) than at other times. For instance, if the Earth in its elliptical orbit is at its farthest point from the Sun (called **aphelion**) and the Moon is at perigee, the Moon will appear relatively larger than if the situation were reversed. If a solar eclipse occurs under these circumstances, the shadow path will be wide. It may span as much as 269 kilometers. On the other hand, if the Moon is near apogee during the eclipse, it may not appear to cover the solar disk completely, even when it is directly in front of the Sun. In this case, the eclipse is said to be **annular** (Figure 3–6).

The duration of a total solar eclipse is determined by how long the eclipse geometry is sustained. This, in turn, is determined by the relative rate of motion of the Sun and Moon and by the velocity of the observer as the particular latitude on which he or she stands rotates with the Earth. Fortunately for the ancients who found the "demise" of the Sun distressing, totality can last no longer than about 7½ minutes.

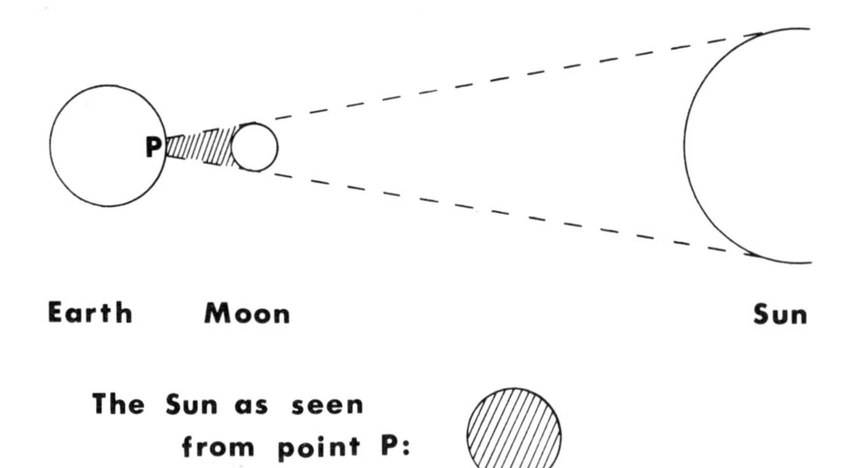

3–6 An annular eclipse. The shadow cone of the Moon does not reach the Earth. An observer watching the Sun from the Earth directly beneath the cone would see a dark disk surrounded by a bright ring, as depicted below. *(Andrea K. Dobson-Hockey)*

A lunar eclipse occurs when the Sun, Earth, and Moon are in line in that order (Figure 3–7). By necessity, then, a lunar eclipse always takes place during a Full Moon, thus making the event even more impressive: A fraction or all of the illuminated disk slowly darkens as the shadow of the Earth covers the Moon. There may be total or partial lunar eclipses. When the Moon passes directly through the middle of the Earth's shadow, the **central** lunar eclipse may last as long as one hour and 40 minutes.

The Moon may not entirely disappear when it is eclipsed. It is still partially lit by sunlight trickling through the Earth's atmosphere. The eclipsed Moon often appears red because red light is more likely to penetrate the atmosphere than blue light. The darkness of this red is governed by cloud systems and high atmospheric particles on the edge of the Moon-facing disk of the Earth.

Solar and lunar eclipses occur with a frequency ratio of about 4:3. However, whereas a lunar eclipse can be observed anywhere on Earth where the Moon can be seen, a total solar eclipse path will pass over a given location only

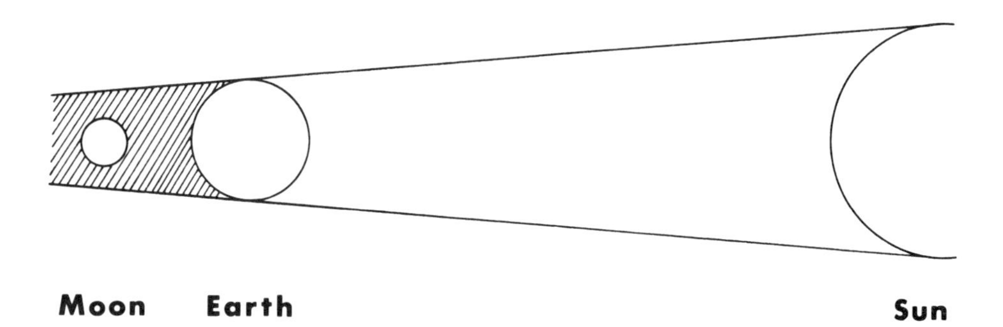

3–7 A lunar eclipse. Because the Moon is smaller than the Earth, the entire lunar disk is in shadow.
(Andrea K. Dobson-Hockey)

once in 300 years of the average. Therefore, a person is more likely to see a lunar eclipse than a solar one. In any eclipse year (12 nodical months), there may be as few as two eclipses or as many as seven (two lunar and five solar or three lunar and four solar).

A certain pattern of eclipses will repeat itself every 223 synodic months. This is the length of time after which the synodic month, nodical month, and anomalistic month are nearly commensurate. After 223 synodic months, the Moon returns to the same position with respect to the Sun, its perigee, and the nodes. This cycle is called the **saros cycle.** Particular eclipse geometries that occur in one such saros will recur in the same order during the next saros. There are approximately 71 solar eclipses per saros. The exact interval of the saros is 6,585.3 days. Because of the fraction, two similar solar eclipses that occur one saros apart will be seen at two different places separated by a third of the distance around the Earth. At every third saros, the same location will experience approximately the same eclipse.

Oh, yes, the answer to the riddle at the beginning of this section is, of course, the Moon—which at one month of age becomes new again. Therefore, it will never be "five weeks old." (The somewhat awkward nomenclature of "old" and "new" Moons was introduced nearly 2,600 years ago by the Athenian statesman, Solon.)

THE CALENDAR

The thing that early people noticed about the Moon, apart from its existence, was that it moved in the sky. Everything that shone in the sky appeared to move in the course of the night, rising in the east and setting in the west, but the Moon also moved with respect to the background of stars. A refinement of this knowledge came with the realization that the Moon follows a *specific* path in the sky and that its motion along this path repeats itself. It was easy to map the path by following the Moon as it moved against the stellar background.

The Moon's position was noted at various times with respect to patterns of stars called **constellations,** which did not seem to change. In this way, the sky was divided up into regions or "houses" in which the Moon dwelt successively as it traveled. To the Babylonians and Greeks, there were 12 such houses, the constellations of the **zodiac.** The Arabs, Chinese, and Indians all used a series of 28 stations or "mansions." In Arabic they were called "alighting places." The constellation patterns that make up these stations are almost identical from culture to culture.

The Moon returns to the same place in the sky after a fixed period. This period corresponds roughly with the Moon's most notable characteristic: its cycle of phases. The idea of the Moon as a time-keeping device may have occurred first to hunters. They needed to know when the Moon would illuminate their path and when it would not. They must have recognized that the Moon appeared one night as a faint sliver in the sky and grew on subsequent nights until it became a bright disk. This disk disappeared in the same manner in which it appeared until it vanished for a few days entirely. It did this over and over again. Furthermore, it always took the same time to go through the cycle. Thus, it became possible to *predict* the phases of the Moon, first night by night and then months or years in advance.

The hunter could now know when the Moon would light his way, and in the process he invented the first calendar.

The day was the first unit of time. The difference that marked it was "as obvious as night and day." However, the Sun was a short-term clock indicating the time of day. People needed a long-term timekeeper, a calendar to keep track of time on a large scale. The changing phases of the Moon were an obvious unit of measure to use for this purpose. The lunar cycle took a length of time that was in step with the pace of primitive life. It was cumbersome to count the number of days between hunts, but it was easy to keep track of the number of "moons" that had passed. Because of the straightforwardness of the lunar calendar, nearly all primitive societies used it.

When people invented agriculture around 8000 B.C., the more complex solar calendar became a necessary advancement. It marked the seasons, which were all-important in determining when to plant crops. In fact, early agrarian societies that still used the month as the unit of their calendar did not always bother to name those months in which agricultural activities did not take place.

The type of calendar used by early societies also depended upon their location on the Earth. In northern or southern latitudes, the difference in the altitude of the Sun between summer and winter was more apparent than it was in the equatorial lands. The seasons themselves were more obvious. People of the temperate zones were more likely to replace the lunar calendar with a solar one. As civilization progressed, however, many adopted both calendars and tried to reconcile them into a **luni-solar calendar.** They did so with varying degrees of success. Still, it is the luni-solar calendar that predominates today.

Aside from the practical benefits of timekeeping, there were religious benefits. Once it was recognized that the heavenly bodies moved in a regular fashion, the practice of periodic worship of the deities associated with them began. Obviously, the gods thought timekeeping was important. Therefore, it was vital to make sure that the sacred days and feasts occurred at the right times. In this way, calendar making became a religious duty. A priest class was often assigned to this important task. Because these holy persons needed to study the sky continually to refine and keep track of the calendar, they were responsible for much of the early astronomical records and were the most knowledgeable individuals in their societies about the workings of the sky.

One continual source of frustration for the calendar makers was the incommensurability of the month and year. In the luni-solar calendar, the unit of the month could not be divided into the length of the year without leaving a remainder of about 11 days. (This remainder is called the **epact.**) The thought that the gods would allow this to be the case plagued the keepers of the calendar with uneasiness and resulted in an ever-intensifying scrutiny of the Moon to find an answer, often resulting in imaginative methods of avoiding the conflict that showed up in early calendars.

The formal calendar first appeared in the Fertile Crescent of the Middle East. Here, clear skies and a pastoral lifestyle allowed plenty of time for studying the movements in the heavens.

The earliest Egyptian calendar was lunar, but lunar astronomy died out in Egypt, leaving as its only remnant the 12-month year. The Egyptians were unusual in that the Moon did not play an important role in their subsequent calendars. Instead, they used the Sun and Sirius, the brightest star in the sky, to predict the critical annual floodings of the Nile.

The Egyptian month was 30 days long. By observing Sirius, Egyptian astronomers knew that the year was approximately 365 days long, not 360, but they made no attempt to reconcile this. Instead, the extra five days at the end of the year were reserved for a special festival. Any transactions completed on these days had no legal status.

One of the first successful luni-solar calendars was designed east of Egypt, in Babylonia. An accurate calendar was very important to the Babylonians. In their astrology, the occurrence of good or bad things rested on whether observations matched the priests' predictions. The detail of Babylonian observations is evident from this astrology. For instance, good fortune would come to the king if a pair of stars set at the moment the Moon reached its highest point in the sky. A new king's reign would be ill-fated if its first month was longer by only a few hours than had previously been calculated. It is no wonder that the Babylonian rulers encouraged observations and calendar improvements.

In the Babylonian calendar, the year was defined to have 12 months. The month began at the precise moment the crescent Moon could be seen in the evening sky and lasted until its next similar appearance. These successive apparitions are called **lunations.** The Babylonians recognized that the year is 365 days long (this was codified by King Nabonassar in 747 B.C.). The difficulty was, as usual, that 12 lunations did not correspond to 365 days. The Babylonians dealt with this by arbitrarily adding a thirteenth month to the year when needed so that the same month would occur in the same season, year after year. This process of adding days or months to a calendar to keep it accurate is called **intercalation.**

When a star can first be seen emerging from the glare of the Sun as it rises in the east, the star is said to be rising **heliacally.** A later test for the time to intercalate called for watching the heliacal risings of certain bright stars. If a star did not rise heliacally in the month it was supposed to, it was time to intercalate a thirteenth month to the current year.

The Babylonians were the first to introduce the arbitrary concept of the week. They made the first day of every month the first day of a new week. Every seventh day was a holiday in honor of Sin. In anticipation of modern blue laws, chariot riding and the cooking of certain foods were forbidden on this day!

The Babylonians finally settled on a mean period of slightly more than 29.5 days for each month. They also measured the length of the year to the minute by averaging repeated observations made over the centuries. Still, they were faced with the problem of intercalation. Remember that in Babylonia the fortunes of the kingdom could rise or fall depending upon whether a New Moon appeared at the precise beginning of the month as it was supposed to. Finally, the Babylonians discovered that 235 lunations occur regularly in the interval of 19 years. A New Moon would occur on the same calendar day after this period of time. This allowed the Babylonians to set up a scheme for intercalating. This cyclical month–year relationship was documented later by the Greek, Meton (fl. 430 B.C.), whence its name, the **Metonic cycle.** Another name for 235 lunations is the "Great Year."

Most peoples of the Middle East adopted or modified the Babylonian calendar. There was one exception. Because the Hebrews were always wary of the influence of other Semitic tribes on their monotheism, they shied away from astronomy and other vestiges of astrolatry. Therefore, it is not surprising that they had one of the worst calendars of their time. Interestingly, it is this calendar that has served in synagogues all over the world for the past 1,600 years.

The Jewish ceremonial calendar began with the year 3761 B.C. It then counted years of 353, 354, or 356 days interspersed with intercalated months. The Hebrew year began with the first or sometimes second lunation after the **vernal equinox** (hence, in Genesis, the expression, "And the evening and the morning were the first day.") After the equinox, 28 days were counted out, and one or two extra days were added from time to time in order to average 29.5 days per month. In approximately A.D. 360, the patriarch Hillel II made some modernizations and put the Jewish calendar into the form that is used today.

The Greeks started a calendar based on the Metonic cycle effective July 16, 443 B.C. By 357 B.C., though, the phases of the Moon were coming one day early. This problem was solved by introducing leap years.

To begin with, the Romans also had a 12-month, 354-day year, with an extra month (called Mercelonius) of 22 or 23 days added every two years. Only 10 of these months were named—January and February came later. The Roman calendar began with 754 B.C., the traditional date for the founding of Rome. Eventually, the calendar degenerated to the point at which intercalation was used discriminatingly to increase the tax year, postpone elections, or lengthen the terms of soldiers' conscriptions. Julius Caesar, in 45 B.C., was responsible for replacing this system with the modern leap-year calendar of 365 or 366 days. Actually, he commissioned Sosigenes of Alexandria (fl. 40 B.C.) to create the Julian calendar. The western world had given up for good trying to regulate the calendar with the phases of the Moon.

In the East, the calendar comes from another tradition. Nomadic Arabs have made ceremonial journeys to cities and oases for millenia, and these gatherings became occasions for exchanging information and goods. The times for these events were set by the Moon. Eventually it was realized that it was preferable for the meetings to be held in the same seasons each year so that the same goods could be traded. Intercalation was borrowed for this purpose. At the end of each gathering, the participants would decide how many months, if any, to add to the year so that they would arrive back at the same time the following year.

The prophet Mohammed considered even this concession to be foreign and returned the Islamic calendar to its lunar roots. The Islamic calendar is the other major lunar calendar in use today. The Koran specifies that the Moon should be used thusly: "Allah has placed the Sun for daylight, the Moon for nightlight and also as an instrument for reckoning time and counting the years." The Islamic calendar starts at the first lunation after the Prophet's flight from Mecca to Medina: Friday, July 16, A.D. 622. It then counts 12 alternating months of 29 or 30 days. There are 354 days in the Moslem year.

Any calendric system that strays from the Moon's synoptic period enough to allow months longer than 29½ days also allows phases of the Moon to repeat during the course of one month. For instance, in any of the months January through December (except February) there can be two Full Moons—if one occurs on the first or second of the month and the other on the thirtieth or thirty-first. This coincidence does not happen often. In fact, the phenomenon's name has become generic: the second Full Moon of the month is the proverbial "blue moon."

PREDICTING ECLIPSES

Perhaps no other astronomical event is as spectacular or, for the uninitiated, terrifying as an eclipse. During an eclipse, one of the great sources of light, the Moon or Sun, is blotted out (Figure 3–8). This occurrence was all the more horrifying to primitive peoples, who had no assurance that the body would return.

On his fourth voyage to the New World, Christopher Columbus told the native Jamaicans that he was responsible for a lunar eclipse. In the midst of it, he threatened to take away the Moon forever if they did not produce food for Columbus and his men.

(Lick Observatory Photograph)

3–8 The Sun in eclipse: September 21, 1922, Wallai, Australia.

Eclipses were almost always assumed to be bad portents. The prophets of the Old Testament invoked eclipses as signs of God's displeasure. Some scholars have explained Isaiah's turning back of the sundial of Ahaz as a partial solar eclipse.

If early biographers are to be believed, eclipses frequently accompany the death of the famous. There are several references to an eclipse on the death of Julius Caesar, though none actually occurred. On the other hand, a lunar eclipse really did take place about the time of King Herod's death. From this we can infer that Jesus of Nazareth was born before March 13, 4 B.C. It is interesting to note that many cultures equate eclipses with the death and resurrection of a god and that New Testament writers describe the darkening of the daylight sky (an eclipse?) at the death of Christ.

Early eclipse watchers thought that during a solar eclipse the Sun and Moon left the sky. The Ojibwa Indian tribe in North America believed that the Sun had simply gone out, and they shot flaming arrows into the sky in an attempt to relight it. Because the Sun always came back, they seemed to be successful.

The Egyptians explained eclipses by saying that the god Apepi stalks the Sun god Ra as he sails the celestial river Urnes. Ra must always be on the lookout for the lurking Apepi. If Apepi succeeds in ambushing Ra, the two meet and an eclipse takes place. In some Egyptian temples, daily Apepi-frustrating services were held.

If an eclipse could be predicted, it would not take people by surprise. Its frightening aspects would be diminished. The one who could predict an eclipse would have great power. No wonder one of the chief goals of the calendar-makers was to be able to predict eclipses.

The Chinese documented a lunar eclipse in 1361 B.C., but the first complete astronomical record of a solar eclipse was made by Babylonian astronomers who noted its beginning and ending times. This eclipse took place on March 19, 721 B.C. The Babylonians' care in making astro-

(Photograph courtesy of Wayne Baggett)

3–9 The Moon in eclipse: July 6, 1982.

nomical observations, the accuracy of which makes them still useful today, explains why they were probably the first to recognize the relationship that would allow them to predict eclipses. It is likely that the Babylonians noticed that lunar eclipses follow a cycle. With this discovery of the saros, lunar eclipses could be predicted. Such predictions were made as early as the seventh century B.C. A lunar eclipse is shown in figure 3–9.

The Babylonians were probably not able to predict solar eclipses. Their observational prowess would have been defeated by the extra dependence of such eclipse observations on the location of the observer. The Babylonians could observe solar eclipses that could be seen only from Babylonia. In fact, they missed, on the average, five out of every six solar eclipses. The sophistication of solar eclipse prediction would have to wait for the Greeks.

The name first associated with prediction of solar eclipses is that of Thales of Miletus (fl. 600 B.C.). The Greek writer Herodotus claims that in 585 B.C. Thales successfully predicted such an eclipse to within a year of its actual occurrence. Supposedly, it was this eclipse, which occurred in the middle of a battle between the Medes and Lydians, that caused both sides to declare peace.

It is understandable that such an eclipse prediction might be attributed to Thales. Thales was the patriarch of Greek science, and an eclipse prediction was a suitable feat for such a figure. A solar eclipse did occur on May 28, 585 B.C. Thales, who traveled widely, may have learned of the saros cycle from the Babylonians or Egyptians, but it is doubtful whether Thales or anyone then alive was capable of anticipating a solar eclipse.

The cause of solar eclipses did not become known until the fifth century B.C. The statesman Pericles was about to embark on a sea voyage when a solar eclipse occurred on August 3, 431 B.C. The sailors aboard the ship regarded this as a bad omen and would not leave port. Pericles, by now familiar with the eclipse mechanism, is said to have asked the vessel's pilot to look at the Sun. He then put his cloak over the man's eyes and asked him if the Sun's disappearance frightened him. When the pilot responded negatively, Pericles pulled the cloak away and said that the Moon's obscuring the Sun was no more an adverse sign than his cloak's obscuring it was. The ship sailed.

The first truly successful prediction of a solar eclipse was probably made by Helicon of Cyzichus (fl. 350 B.C.), using the saros cycle. It was a major achievement for its time.

Hipparchus of Nicaea (fl. 140 B.C.) contributed

much to the theory of eclipses. He discovered the inclination of the Moon's orbit to the ecliptic and the orbit's eccentricity. He defined the Line of Apsides and Line of Nodes, which enabled him to time the different months (synodic, siderial, nodical, and anomalistic) precisely by using ancient records. Armed with this information, Hipparchus was able to explain the saros cycle and became one of the most accurate eclipse predictors to that time. In fact, he was able to fix the time of lunar eclipses to within two hours.

By the first millenium, eclipse prediction was routine. When an eclipse was predicted to occur on the birthday of the Emperor Claudius, he announced it well in advance so that people would not become uneasy. Presumably he also wished to thwart anyone who wanted to validate the omen by plotting his assassination. An indicator of the backsliding that occurred during the Middle Ages was another emperor, Louis of Bavaria, who was less enlightened. He supposedly died out of sheer terror during an eclipse in A.D. 840, thereby precipitating the breakup of the Carolingian empire.

The Chinese had an intense interest in eclipses, though originally the Moon was not thought to be involved. It was feared that eclipses were caused by a dragon devouring the Sun. Aristocrats and peasants alike would join in, clanging gongs and setting off firecrackers in order to scare away the beast with their noise. Once again, this method of ending a solar eclipse appeared to be successful because the Sun was always "regurgitated" in its original form.

Trying to predict eclipses was a major function of the priestly class in China. Legend has it that in 2136 B.C., the court astronomers Ho and Hsi were executed for paying too much attention to Earthly delights and ignoring a solar eclipse. It is almost certain that Ho and Hsi did not exist and that the ancient Chinese were never able to predict solar eclipses accurately. Astronomy was practically abandoned after the fifth century B.C. in China. In fact, the Chinese calendar fell into such a state of disarray after successive "improvements" made by emperors wishing to leave their mark on it that when in A.D. 1610 a solar eclipse was missed—even though the technique for predicting it was by then long known in Europe—the emperor fired his calendar keepers and turned the task over to European Jesuit missionaries living in China at the time.

THE MOON IN THE UNIVERSE

In his novel, *2001: A Space Odyssey,* Arthur C. Clarke introduces us to the character of Moon-Watcher, a Pliocene hominid living in 3,000,000 B.C. Moon-Watcher is so named because he notices the Moon and wants to touch it. Frustrated in his attempts as a child, in adulthood he acknowledges his early foolishness and realizes that he must first find a tree of proper height before he can reach the Moon.

Moon-Watcher, or at least someone like him, was the first lunar astronomer. Moon-watcher represents the first being on Earth who recognized the Moon as a distinct entity in his environment. In this way, the study of the Moon began. In fact, Moon-Watcher unknowingly touched upon several of the questions that would perplex astronomers for eons: How far away is this shiny disk, and how large is it?

Geometry eventually established that to answer one of these questions one had to answer the other, but at first all that was certain were the relative positions of things in the sky, not their absolute distances. For instance, the Moon must be closer to us than the Sun and the stars because it can occult both. Less conclusively, the Moon moves more rapidly than any other planet so it is probably closer to us than are the planets.

Early Mediterranean people thought that the Moon rose out of the eastern sea, drifted overhead, and then set into the western sea. Thereupon, it floated back to its starting place. To the Babylonians, the sky was solid, and the Moon traveled across it by entering through one door in the east and exiting through another in the west. The Egyptians thought that the heavens were a sort of box covering the Earth. The box was rectangular, and the Moon in its journey had to make a right-angle turn whenever it came to a corner!

None of these theories explained the visible universe very well.

Most of our early correct concepts of the universe were the work of a remarkable string of Greek thinkers who flourished during Greece's Golden Age between 600 and 300 B.C. At that time, Greece was the center of technical and scientific knowledge.

Travelers not only brought wealth to Greece but also information from all over the known world. They brought the knowledge that when one is traveling the Moon appears to shift in its position very little against the background of stars just as a distant tree appears to move only slightly against more distant mountains when one steps from side to side. The Greeks brought to bear their powers of synthesis and later invoked geometry to use this argument of **parallax** in answering the first question. The Moon (and Sun) must be very far away indeed.

We have already encountered Thales, the founder of Greek science. Attributed to Thales is the knowledge that the Moon shines because of reflected light from the Sun and that the Moon becomes invisible when in conjunction with the Sun because it is a dark body and is hidden in the Sun's glare.

Thales' most important contribution to astronomy, though, was the founding of the Ionian school of fellow academicians, students from whom flowed many major contributions to the study of astronomy. One of these was Anaximander fl. 580 B.C.). Anaximander assembled a picture of the universe as he envisioned it. Earlier Greeks had only alluded vaguely to a celestial dome with the Sun, Moon, and Earth hanging from strings or mounted on studs projecting from it. To Anaximander, the Moon and Sun were really holes in an opaque wheel through which the light of a great fire shone. Anaximander's universe, although odd, was a first attempt to explain real phenomena.

Meanwhile, another school of Greek thought was founded by Pythagoras of Samos (c. 580–500 B.C.). Pythagoras himself was the geometer of Pythagorean Theorem fame. Pythagoras also turned his attention to the Moon, though it is difficult to separate his ideas from those of his followers.

In the Pythagorean universe, all bodies orbited a great central fire in concentric circles. The ratios of the distances of these orbits to one another were determined somewhat mystically by musical intervals, the so-called Harmony of the Spheres. Some Pythagoreans let a revolving Earth cause the daily rising and setting of the Sun, Moon, planets, and stars.

Foremost among the early Greek selenographers was Anaxagoras of Clazomenae (c. 500–430 B.C.). Anaxagoras was a disciple of Thales. It was he who first explained eclipses. He also said that the Moon was as large as the Peloponnesus (about the size of Massachusetts), considered to be a tremendous area at that time. Anaxagoras believed that the Moon and planets were stony like the Earth. Specifically, the Moon was a flat disk with plains and ravines. In fact, he thought, the Moon might even be inhabitable.

Although learned persons of his time did not believe in the divinity of the Moon, Anaxagoras happened to publish his theories during a religious revival. For his accomplishments, Anaxagoras was sentenced to death. Only the political intervention of a friend got him off with mere permanent exile.

Eudoxus of Cnidus (c. 400–350 B.C.), a student of the philosopher Plato, envisioned a universe in which the celestial bodies were mounted on concentric, crystalline spheres with the Earth in the center. Circular motion was caused by the turning of these spheres. Because the planets did not appear to move uniformly in the sky, Eudoxus gave them four spheres each. These four spheres turned within one another. By using multiple spheres for each body and linking them together, Eudoxus was able to better model their motion as seen from the Earth.

Aristotle, also a student of Plato, has had the greatest effect on people's perceived nature of the Moon because it was his ideas that were codified and held intractable for more than a millenium. Aristotle (Figure 3–10) offered as a proof of the spherical Moon its crescent shape before its first quarter and after its third quarter, with the cusps always being pointed away from the Sun. Likewise, he argued that the Earth is spherical because the sphere is the only shape that will project a circular shadow onto the Moon during a lunar eclipse, no matter where the Moon is in the sky.

(Photograph courtesy of the AIP Niels Bohr Library)

3–10 Aristotle.

Aristotle's universe was an embellishment of that of Eudoxus, but whereas Eudoxus' celestial spheres were mathematical models, Aristotle's were real. Aristotle believed that all the spheres were driven by the motion of the outermost shell, which held the firmament. This sphere was the **primum mobile.** Aristotle introduced **reacting spheres** between those that had hitherto moved independently. Now all the spheres interacted. By using 55 interlocking spheres, Aristotle was able to approximate the motion of all the objects in the sky. Again, the Earth was at the center of the spheres.

Aristarchus of Samos (fl. 270 B.C.) elegantly approached the problem of the sizes of and distances to the Moon and the Sun. Aristarchus realized that a line drawn from the Earth to the Moon and a line drawn from the Moon to the Sun are at right angles to each other when the Moon is exactly at its first or third quarter. The Sun is not exactly 90 degrees away from the Moon in the sky at this time; it is slightly less. By using the geometry of Pythagoras, Aristarchus was able to estimate the relative distances to the Moon and the Sun. By estimating their apparent diameters, Aristarchus was then able to assign sizes to these bodies.

There were two flaws in Aristarchus' project. One was that he estimated the angular separation between the Moon and the Sun to be 87 degrees; the actual value is much closer to 90 degrees. The other flaw was that he estimated their apparent diameters to be 2 degrees, four times too big. Hence, his estimates of distance were too small. The angular separation measurement was beyond Aristarchus' capability. Determining the exact moment of the First or the Third Quarter is difficult even today. The measurement of apparent diameter, though, is easily made, and Aristarchus' inaccuracy points out the handicap of Greek science: It was stong on theory but weak on experimentation.

Aristarchus also speculated on the concept of a **heliocentric** universe rather than a **geocentric** one. That is, he suggested that the heavenly bodies might be moving on crystalline spheres centered on the Sun rather than on the Earth. Only the **sublunar** sphere traveled around the Earth, which was on a heliocentric sphere itself. The outermost sphere, which held the stars, was vastly greater in volume than the others. This theory was not taken seriously, and the heliocentric theory was never mentioned again in Greek science after Aristarchus.

In the later centuries of the Greek era, there was a switch in Greek science from philosophical theorizing about nature to observing and cataloging it. The later Greek astronomers are exemplified by Hipparchus (recall his work on lunar eclipses). Hipparchus believed that if one truly wants to understand the universe, one must measure it precisely first. His accomplishments did not represent giant leaps in thought but, rather, steady, scientific progress.

Hipparchus constructed a method for determining the relative distances to the Moon and Sun during lunar eclipses by using his own invention, trigonometry. He concluded that they were much farther away than had previously been supposed. Hipparchus also introduced the idea of eccentric, circular planetary orbits that were not centered on the Earth in order to explain their apparent lack of uniform motion.

Greek astronomy was summarized by Claudius Ptolemaeus (fl. A.D. 140), better known as Ptolemy. We know of Hipparchus' work through Ptolemy. An astronomer in his own right, Ptolemy tried to measure the absolute distance to the Moon by measuring its altitude above the horizon at successive lunations. He also discovered an irregularity in the Moon's motion called **evection.** However, Ptolemy is best known for recording the state of astronomical knowledge in

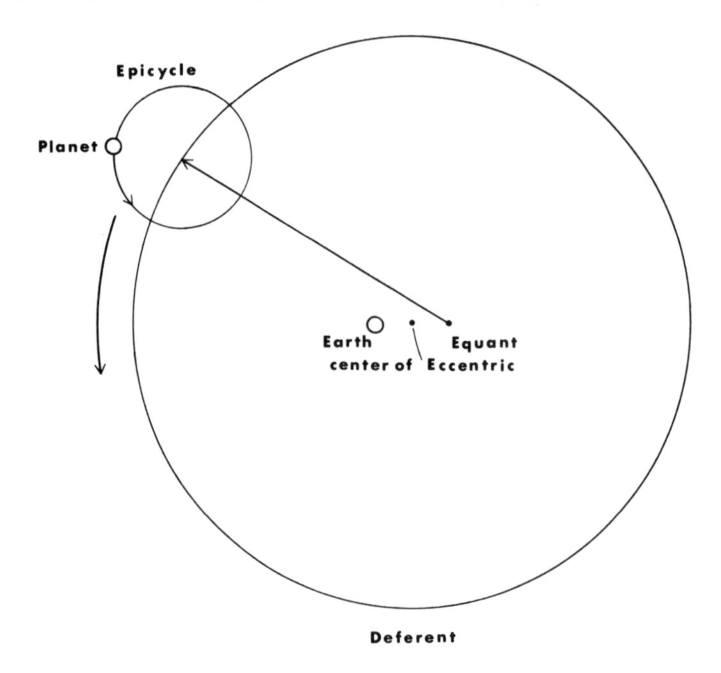

3–11 The Ptolemaic system. A representation of the motion of a hypothetical planet in shown.
(Andrea K. Dobson-Hockey)

his time, published largely in his tome, *The Almagest,* a title given to it by later scholars. Into *The Almagest* went a model of the universe based on that of Hipparchus.

Ptolemy used a geometrical device called an **epicycle,** first described by Apollonius of Perge around 230 B.C., to explain celestial motions (Figure 3–11). In the epicycle universe, each planet could be imagined to be fixed to the circumference of a rotating circle, the epicycle, the center of which was attached to the circumference of a greater circle, the **deferent.** The Moon traveled on a circle that was centered a little to one side of the middle of the Earth, hence the eccentricity of its orbit. The same held true for the Sun. Objects underwent uniform *angular* motion with respect to a point called the **equant.** This complex model required 70 separate motions, but it did greatly facilitate the prediction of the positions of celestial bodies.

Ptolemy's work was dogma by the time serious astronomy was once again practiced in Europe.

4
THE FACE OF THE MOON

THE RENAISSANCE MOON

Although the first millenium A.D. was neither as ignorant nor as "dark" as many believe, it is true that very little intellectual light shone on the Moon. In astronomy, maintaining the status quo was difficult enough. We owe many of the records that survived antiquity to Moslem scholars and Christian monks who carefully perpetuated the works of Aristotle and Ptolemy by caring for and continuously copying their manuscripts. It is unsettling to think that at many points in history fire or some other calamity could easily have made the names and accomplishments of the scientists and philosophers we met in Chapter 3 unknown to us. The printing press helped, but only today after the information deluge of the twentieth century can we be assured that, barring the ultimate holocaust, the knowledge of the millenia will survive—on tape, floppy disk, or in print—and will never be lost.

This diligence came at a price, however. As the works of the Golden Age of Greece persisted, they also gained importance. They transcended their authors' intent to the point at which they were considered to be synonomous with Truth. Aristotle especially was held inviolate, ironically by a church that considered him pagan. Although Ptolemy's theories were the basis for calculating the positions of the Moon, Sun, and planets, the Aristotelian spheres were thought to be closer to reality. Aristotle's ideas on the structure of the solar system and the nature of motion would become particularly troublesome when humankind once again grappled seriously with understanding the laws of nature.

The progress that was made in lunar science during these years took place in the Near and Middle East. There, scientists kept records on the positions of the Moon and planets during a scientific revival that started in the seventh century. Astronomical measurements using the **astrolabe** and other instruments were a welcome change from the abstractions of early Greek science.

Ahmed ben Muhammed al Fargani (fl. 800) compiled a table for the motion of the Moon and Sun. Muhammed al Battani (858–929) published records on the parallax of the Moon and on eclipses. In the late tenth century, Muhammed Abu'l Wefa al Bûzjani (959–998), a Persian, translated and assembled the parts of *The Almagest* and may have discovered an irregularity in the Moon's motion called **variation.** These data would become the numbers used in equations of celestial motion during the second millenium A.D. when mathematics would be applied to astronomy.

It is somehow not surprising that the first addition to lunar knowledge to come again from Europe was the work of the man whose name is nearly synonomous with the Renaissance: Leonardo da Vinci (1452–1519). Da Vinci's contribution concerned a phenomenon called the "old Moon in the young Moon's arms." During the New Moon, the outline of the Moon can often still be made out as a faintly illuminated disk. Like most everything concerned with the Moon, this display was considered a portent of disaster. The Greek, Poseidonius, had explained it as light passing through the semitransparent Moon. Others had said that the Moon shone faintly by itself. Da Vinci recognized the obvious light source: the Earth. Sunlight reflecting from the daylight hemisphere of Earth is again reflected off the Moon. It is a tiny amount of light but enough to contrast with the blackness of space. **Earthshine** can also be seen filling out the lunar disk during the crescent or gibbous phases, but it is greatly obscured by the brightly lit portion.

The Renaissance expanded beyond Italy. It gave us the greatest Polish astronomer of all time: Mikolaj Kopernigk (1473–1543). We know him as Copernicus. (It was customary for European scientists of the time to Latinize their names.) His book, *De Revolutionibus Orbium Coelestium* (1543), so radically countered established ideology about the organization of the universe that Copernicus dared publish it only on his deathbed, and it remained on the notorious *Index Librorum Prohibitorum* until the nineteenth century. Much is misunderstood about Copernicus' heliocentric theory. (Figure 4–1). He did not in one fell swoop properly set up the solar system. He did not overturn the system of epicycles derived by Ptolemy. His goal was simply to make this system less cumbersome by making one alteration: The Earth now traveled on a deferent around the Sun rather than the other way around. So did the other planets. The

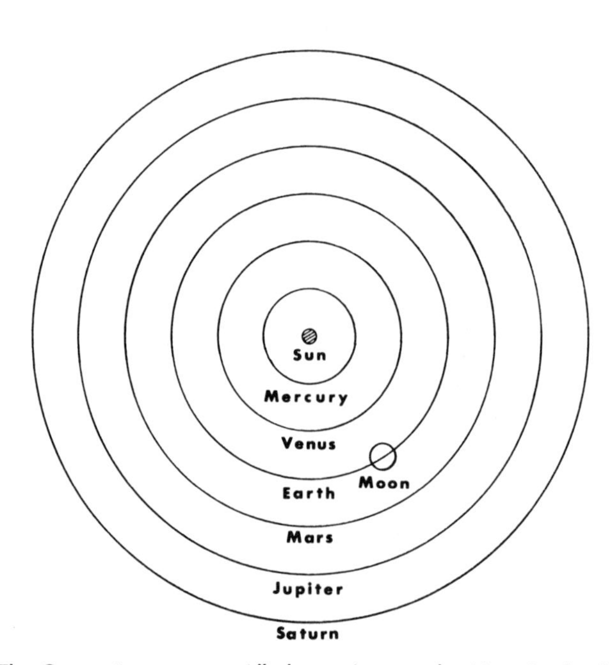

4–1 The Copernican system. All planets (except the Moon) orbit the Sun.
(Andrea K. Dobson-Hockey)

Moon stayed where it was, and the stars remained fixed on an outermost sphere. However, fewer epicycles were now necessary to explain the motions of the heavens. This refinement might have been welcomed had it not required moving the Earth from the center of the universe and therefore been labeled heretical.

Copernicus' work was conservative by some standards, but it was the first step toward the de-emphasization of the Earth in astronomy and a truer, less biased picture of the cosmos.

Observations were still necessary to flesh out the intricacies of planetary motion, and the foremost of the pretelescopic observers was Tycho Brahe (1546–1601) of Denmark. The young Tycho first watched a solar eclipse that took place on August 21, 1560. It so impressed him that he decided then and there to study astronomy. His ultimate goal was none other than an explanation of all planetary motion once and for all.

In 1566 Tycho won notoriety by supposedly predicting that an eclipse foretold the death of the Sultan of Turkey. The Sultan did die. It was simply not well publicized that he had met his death considerably before the eclipse had taken place.

Tycho got his big chance to measure the universe when King Frederick II of Denmark offered him the island of Hven, near Copenhagen, in 1576. On Hven, Tycho built the first observatory worthy of the name and called it Uraniborg. Uraniborg housed both Tycho and the most sophisticated astronomical instruments of the age. There, he collected a vast store of astronomical data.

Tycho discovered two irregularities in the Moon's motion: variation and the "annual inequality." Like Arab measurers before him, Tycho's discoveries required that more epicycles be added to Ptolemy's system. The epicycle theory was clearly on its way to collapsing under its own weight. Tycho approved of the simplifying aspect of Copernicus' theory, but he was hesitant to let the Earth move. His compromise was the **Tychonic System** (Figure 4–2) in

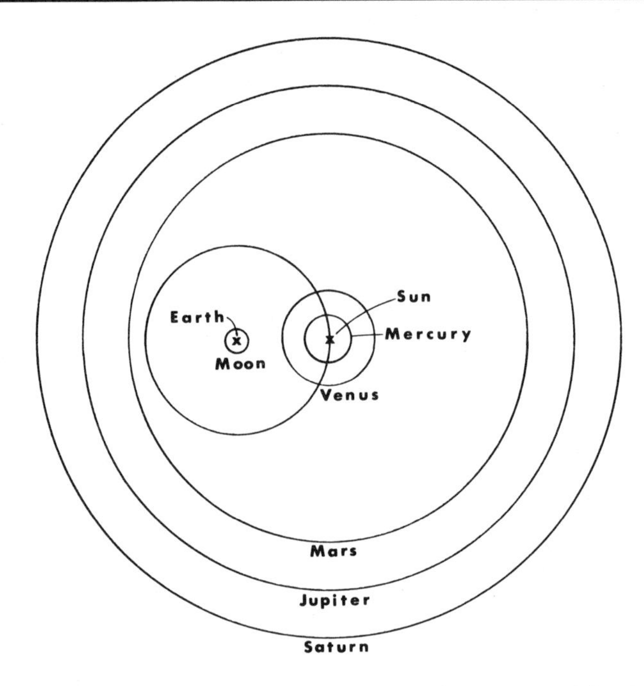

4–2 The Tychonic system. All planets (except the Moon and the Earth) orbit the Sun. The Moon and the Sun orbit the Earth.
(Andrea K. Dobson-Hockey)

which all the planets orbited the Sun, but the Sun itself still orbited the Earth (along with the Moon), thus sidestepping the theological problems. Tycho realized that there needed to be proof that his or Copernicus' system worked better than any other. On his deathbed he charged his assistant with developing this proof.

Tycho's assistant, whom he commissioned to prove his theory, was a good choice for the job. He was Johannes Kepler (1571–1630), a German mathematician who had the computational skills necessary to try to match theory with observation. Kepler was an astronomer and reticent astrologer by vocation. He was the first to explain the dim illumination of the Moon during a lunar eclipse as coming from sunlight passing through the Earth's atmosphere. After a few months with Tycho, Kepler found himself heir to the great observer's data. Kepler was an avid Copernican and didn't spend much time with the Tychonic System. However, he still could not reconcile the heliocentric system with reality. Finally, he left the "perfection" of the circle entirely and sought a new orbital shape, the ellipse, on which to base the solar system. It worked! Kepler was able to describe the motions that Tycho had seen by using the ellipse, and the epicycles were gone for good.

The ellipse is the locus of points the sum of whose distances from two other points, called focii, remains constant (see Figure 4–3). **Kepler** set the Sun at one focus of an ellipse and then set the planets orbiting about it in elliptical orbits. Elliptical orbits are the first of Kepler's famous three **Laws of Planetary Motion.**

Remember the trouble that the epicyclists had with the rate at which the planets move? It seemed logical that the planets should move uniformly (that is, their speed should remain constant), but observations showed that this was clearly not the case. Kepler saved uniform motion by replacing it with his Second Law, which said that in its elliptical plane the line connecting the Sun and a planet sweeps out equal areas in equal amounts of time (Figure 4–4). Uniform area replaced uniform motion.

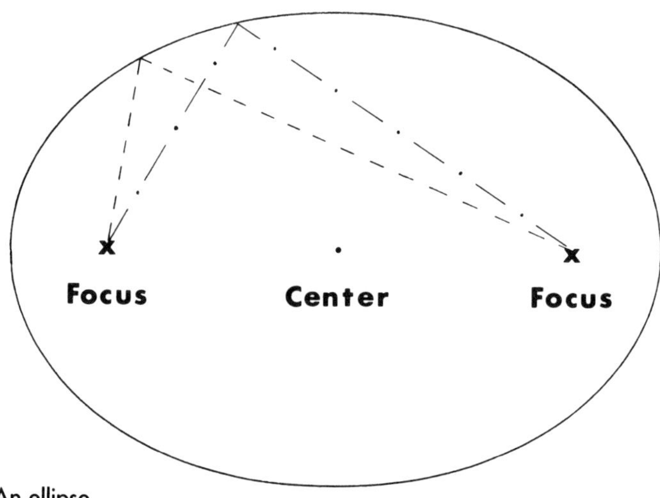

4–3 An ellipse.
(Andrea K. Dobson-Hockey)

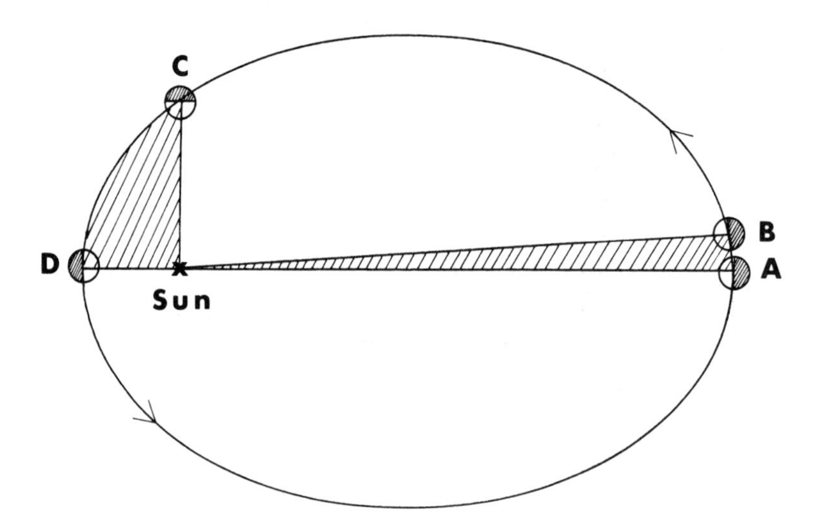

4–4 Kepler's Second Law. A planet will travel between points A and B and points C and D in its orbit in equal periods of time. The hatched "swept out" regions are of equal area.
(Andrea K. Dobson-Hockey)

Kepler's Third Law says that the square of the period of revolution of a planet is proportional to the cube of its distance from the Sun:

$$P^2 \alpha r^3$$

This is the golden rule that controls planetary motion. It allowed relative determination of planetary distances for the first time.

Let's see what the implications of Kepler's laws are for the Moon. First, the Earth now replaces the Sun as one of the focii. The Moon orbits the Earth (more precisely, the barycenter) in an elliptical orbit. When it is closest to the Earth, it is at perigee. When it is closest to the other focus of the ellipse, it is at apogee. From the Second Law we find that for the Moon to sweep out equal areas in equal times it must move faster in its orbit near perigee, when the distance between the Earth and the Moon is smallest, than it does at apogee, when the distance between the Earth and the Moon is larger.

The Third Law would become useful for determining the relative distances of the Moon and Sun once a proportionality constant (governed by the masses of the Earth and Sun) could be calculated:

$$\frac{P^2 \text{ earth}}{r^3 \text{ earth}} = \frac{kP^2 \text{ moon}}{r^3 \text{ moon}}$$

Kepler was extremely successful in explaining the motion of the planets. In the case of the Moon, though, the foregoing application of Kepler's Laws worked only approximately. There were still deviations of the motion of the Moon that he could not explain. This no doubt perplexed Kepler, but he may have been relieved to know that although it would take centuries still to work out the complexities and minor perturbations affecting orbiting bodies, no one would again attempt to do so using epicycles. They would all begin with Kepler's Laws.

THE TELESCOPE

Until now we have considered the Moon only as it moves in space. For all practical purposes our discussion could have dealt with a featureless point. For much of history people preferred to think of the Moon as just this. After all, the heavens were a place of perfection in contrast to the obvious imperfections of our own world. The Sun and the planets were pure disks or points of light. These eternal featureless orbs were not in anyway marred or blemished. The Moon, though, was a problem. Its shiny face seemed to have markings on it—patches of light and dark. This was unacceptable and accounts for much of the suspicion of the Moon in myth and religion. A way around this problem was to ignore it, and so the Moon's face went unconsidered by lunar science.

Yet there were those who chose to believe their eyes or who did not know that there were not supposed to be markings on the Moon. Many primitive peoples saw patterns there. They were drawn to making pictures from the lunar face much in the same way in which they recognized patterns in the stars and called them constellations. They saw a dog or a fox or a giant on the Moon. In southwestern North America it was a coyote. The Chinese saw a three-legged toad or the complicated scene of a hare sitting in front of a bowl while it pounds rice with a pestle held in its paws.

As has been mentioned previously, rabbits were popular. The Moon hare was seen by such diverse groups as Hindus, Tibetans, Amerinds, and the people in Zululand.

Humans appeared as well. There was the Old Lady or the Old Man. Alternately, there was a child or a Moon Girl. The Greeks and Romans chose to see a female face on their Moon goddess. Today, many of us in the west opt for the Man in the Moon. Shakespeare's verse, "This man with lanthorn, dog, and bush of thorn" in *A Midsummer Night's Dream,* Act V, Scene I, is a reference to him.

Although there were those persons who saw all sorts of creatures and scenes in the Moon, there were also a few who saw landscapes. Recall the Anaxagoras believed that the Moon was covered by plains and ravines. Aristotle, though, said otherwise. The Moon was smooth, self-luminous, and perfect. He conceded that the Moon might be subject to some imperfection because the sphere on which it was mounted was adjacent to the imperfect Earth, but he thought it more likely that the lunar markings were merely reflections of irregular Earth features in its mirror-like surface.

Others said that lunar features were transient phenomena of our own atmosphere. This theory was weak from its inception because the patient observer could tell that from one Full Moon to the next the lunar markings remained in the same places. We now know that this is because the Moon rotates in the same period of time that it orbits the Earth; thus, it always presents the same side to us. When we speak of the face of the Moon, we are talking about the **lunar Nearside.** The other side of the Moon is the **lunar Farside** and is forever turned away from us on the Earth.

Plutarch (c. 46–119), in his *De Facie in Orbe Lunae (On the Face of the Moon)* concluded that the Moon does have markings and that they are intrinsic. He discussed mountains and other Earth-like features of its terrain. Aristotle was heard; Plutarch was not. The Renaissance still thought the Moon to be a perfect disk, but by now it was realized that the markings on the Moon remained unchanging and needed to be explained.

The maddening thing was that the markings were not distinct enough! For instance, if they were reflected Earth-features, they were unrecognizable. Perhaps the atmosphere distorted them. A closer look was needed.

The man who provided that look was the third of a great trio of astronomers working near the turn of the seventeenth century. He was a contemporary of Kepler named Galileo Galilei (1564–1642). Like Tycho, his first name suffices to place him, so great were his contributions.

Galileo was an Italian scientist, many of whose accomplishments came under the patronage of the Grand Duke of Tuscany. Before becoming one of the greatest astonomers of all time, he became one of the greatest physicists of all time with his experiments on motion. These experiments lie outside the scope of our story. When we meet Galileo, it is 1609. He has learned of Dutch experiments with placing two glass lenses in one's line of sight in order to magnify an object, and he is about to construct such an apparatus for himself.

Galileo was the first to point the new invention at the night sky, although another early observer of the Moon, Simon Mayer (Marius, 1573–1624), also claimed this distinction. Almost at once, he discovered sunspots, the phases of Venus, the rings of Saturn, and the satellites of Jupiter. However, no view was more spectacular than that which awaited Galileo when he looked through his instrument at the Moon. Indeed, the Moon may have been the first astronomical object ever seen through a **telescope.**

What Galileo saw was not a disk but a world that stood out in high relief. He saw mountains and valleys there just like those on the Earth. He saw shadows everywhere. Peaks near the **terminator** (the line dividing light from dark) stood with their tops illuminated and their bases in shadow, testifying to the texture of the Moon's surface. Galileo even used this phenomenon to measure the height of lunar mountains and found that they compare favorably with their counterparts on Earth.

People scoffed at Galileo's rough Moon. They refused to look through his "spyglass" (Galileo's term), or when they did, they said that they couldn't see anything. (To be sure, one had to have good eyesight and be dexterous to get much use out of Galileo's rather crude instrument!) Others tried to rationalize what they had seen. Perhaps the rough features were imbedded in a smooth crystal as in a glass paperweight. This unseen sphere was what counted, they said, and it was smooth and perfect. An irritated Galileo replied that if someone insisted on constructing such an invisible smooth sphere over the Moon in their imagination, then he, Galileo, would place invisible rough mountains on top of it! The perfect lunar orb was gone for good, but the imperfect neighboring world that the Moon had become was much more fascinating.

Although Galileo saw a great deal on the Moon that was familiar to him, there were also strange new features. For instance, there were those dark patches. The telescope revealed these to be relatively smooth compared to the lighter, mountainous areas. Furthermore, the boundaries of these regions were abrupt. Galileo chose to interpret them as shores that separated the light land on the Moon from the dark lunar "seas."

In his writings, Galileo referred to his seas (in Latin, **maria**) as "ancient spots," that is, those dark areas that had always been visible—albeit vaguely—without a telescope. He did so to differentiate them from the smaller, more numerous spots he could now see through the telescope. These circular features were very un-Earth-like but could be seen all over the rough parts of the Moon. They were definitely topographical features because their rims cast shadows. Galileo said that they reminded him of the eyes of peacock tails. Today, we call them, in less colorful prose, the lunar **craters.**

Galileo published his observations in 1610 as the *Siderius Nuncius (Messenger from the Skies).* Although his description of the lunar surface was controversial, his discovery of Jupiter's satellites was clearly heretical. Here were bodies that orbited a parent body other than the Earth. If they could do so, why couldn't the planets? The discovery of the Galilean satellites pointed away from Aristotle and toward Copernicus, whose theory Galileo embraced.

Publication of the *Siderius Nuncius* was at the same time a monumental advance in astronomy and a violation of the law. A lunar surface much like the Earth's, satellites around Jupiter, and spots on the Sun were considered heretical because they violated Aristotelian geocentrism and the then-accepted picture of the divine nature of the heavens. Ironically, Italy, which had conceived the Renaissance and its spirit of toleration, now censored Galileo. His book, like many of the great early works on astronomy, was banned, and Galileo (later to become blind because of his zeal in watching the Sun through his lenses) was forced to recant and spend the rest of his life virtually imprisoned.

From the time of Galileo to the landing of the first spacecraft on the lunar surface, what was learned about the Moon was learned using the telescope. What was this new device that sealed Galileo's fate and opened up the universe for inspection? It seems to have been invented by Hans Lippershey of Holland and others early in the first decade of the sixteenth century. It consisted of two lenses held together in a tube. When one looked through it, distant objects appeared nearer and small objects appeared larger. It was an interesting novelty and useful for such things as looking at ships far off to sea. Its military value as a reconnaissance device was, of course, not overlooked. By 1609 telescopes could be purchased in Paris. Galileo himself made about 100 of them and gave them to prominent people all over Europe.

How does the telescope allow one to look far away? The telescope relies on a property of light called **refraction** (Figure 4–5). When a light ray travels through a medium (say, the air) and intersects another medium (say, glass) at an angle, the ray changes its direction of travel. A lens is simply a piece of glass ground so that light rays that are traveling parallel to the axis of the lens will be bent (refracted) as they enter it and are all directed to a common point. The lens is said to focus the light at this point.

Imagine a source of light, perhaps a star, that is radiating in all directions but is also very far away. By the time its light reaches us, the light rays are nearly parallel. The rays that happen to strike the first lens of a telescope, called the **objective,** are brought to a focus at a point (the **focal point**) behind the lens inside the telescope. The distance between the center of the lens and the focal point is

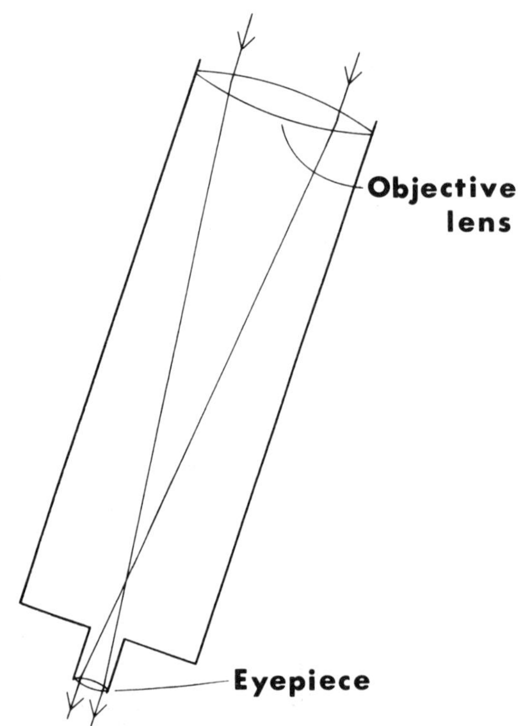

4–5 A refracting telescope. The two lines entering the telescope represent rays of light coming from a point source very far away.
(Andrea K. Dobson-Hockey)

called the **focal length** of the lens. Now suppose we put another lens behind the focal point at a distance equal to its focal length. The opposite effect occurs. The rays of light diverge past the focal point, enter the second lens, and are refracted so that they are parallel again. If we put our eye behind the second lens, we will see an image of the star. For this reason this second lens is called the **eyepiece** of the telescope. Kepler made fundamental improvements on the telescope eyepiece and outlined all of elementary optics in *Dioptrice,* published in 1611.

What is the advantage of a telescope? Obviously, we don't need a telescope to produce an image of a star (which even in the largest of telescopes will remain a mere point of light). Our own eyes can do that just by looking up at the sky. However, an eye will receive only the light from the star that is intercepted by the pupil, which is, at most, an opening of about 1/4 inch in diameter. Because the eye does not store light, at any given time we see only as much light as comes through this quarter-inch aperture. The advantage of the telescope is that the diameter of the objective lens can be much greater. When we look at the star image through a telescope, we receive all the light gathered by the objective. A 12-inch telescope (one that has an objective 12 inches in diameter) can be likened to a bucket set out in the rain. Compared to the telescope, our naked eye is like a thimble set next the bucket. Clearly, in a given length of time, the bucket will collect more rain water than will the thimble. Similarly, the 12-inch telescope will collect more light than our unaided eye. Therefore, when we look at a star or anything else through a telescope, it appears brighter.

So far, we have not cited an advantage of looking at the Moon through a telescope. The Moon is not a faint object that needs to be made brighter. In fact, in a large aperture telescope, the Moon may be uncomfortably bright to our eye, making it difficult to look at. To find the advantage of viewing an *extended* object such as the Moon through a telescope, we look at another property of the telescope, its focal length.

Imagine that now we are pointing our telescope at the Moon. Consider the light that is coming to us from a point on its surface near the northern **limb** (the northern edge of the lunar disk). The Moon is far away, so the light rays from this point arrive at our telescope nearly parallel. They enter the objective and are brought to a focus. Now consider a similar point on the Moon near the southern limb. Parallel light rays from that point come toward the telescope from a slightly different direction. They too are focused by the lens but at a point above that of the first point. Every point in the disk of the Moon is similarly focused. In this way, an image of the Moon is built up (upside down) behind the objective in the **focal plane** (Figure 4–6). Now the light rays diverge past the focal plane and enter the eyepiece, which produces an image made up of parallel light rays. An observer behind the eyepiece now sees an image bigger than that viewed without the telescope. That is, the apparent angular size of the image is greater than that of the Moon as we see it unaided in the sky. The telescope has **magnified** the Moon.

The magnification, or power, of a telescope is simply the ratio of the angular size of an object viewed with and without the telescope. It is also the ratio of the focal length of the objective to that of the eyepiece. Although the focal length of the eyepiece can be only so short for practical reasons, in theory a telescope can be any length so that the focal length of the objective can be very long. Does this mean we can have a telescope with unlimited magnification? Remember that, whatever the image size, the amount of light available to create it is determined by the size of the objective. A large image has to spread the light out over a greater area in the focal plane. Eventually, the image becomes too faint to be of any use. Of course, one could overcome this problem by using a larger objective, but there are technical limits to doing this *ad infinitum.*

More important than magnification is **resolution.** When the aperture of a telescope is increased, the eye behind it can separate adjacent points in the lunar image that

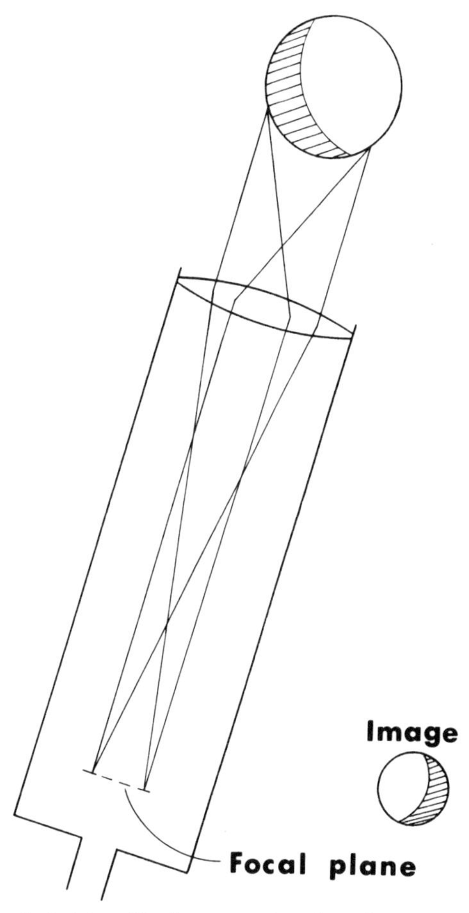

4–6 An extended object. The lines entering the telescope represent rays of light coming from the two limbs of the Moon. The image at the focal plane is depicted in the corner of the diagram.
(Andrea K. Dobson-Hockey)

were previously indistinguishable as distinct features. More detail becomes apparent. It is this added resolution that gives us a better look at the Moon and reveals details that were hidden before the invention of the telescope.

The ultimate limit to resolution is a function of the diameter of the aperture (another reason for a large objective), but the limit is seldom achieved in practice. This is because the blurring and distorting effects of the Earth's atmosphere impose a practical ceiling on the resolution obtainable from Earth-based telescopes.

The telescope that has been described so far is of the type constructed by Galileo. It is specifically a **refracting telescope** (refractor) and consists of two or more lenses mounted with their axes in line. A completely different type of telescope was invented by Isaac Newton (whose still greater accomplishments appear in the next chapter). The **reflecting telescope** (reflector) uses a curved mirror as an objective. Parallel light rays strike the spherical or parabolic mirror and are reflected to a focus. The reflector works in much the same way as the refractor except that its focal plane lies in front of the objective. In a **Newtonian telescope** a small flat mirror redirects the image from the focal plane off to the side and the eyepiece (Figure 4–7). In a **Cassegrain** reflector, the light is directed back through a hole in the objective mirror and out the back through an eyepiece mounted behind it.

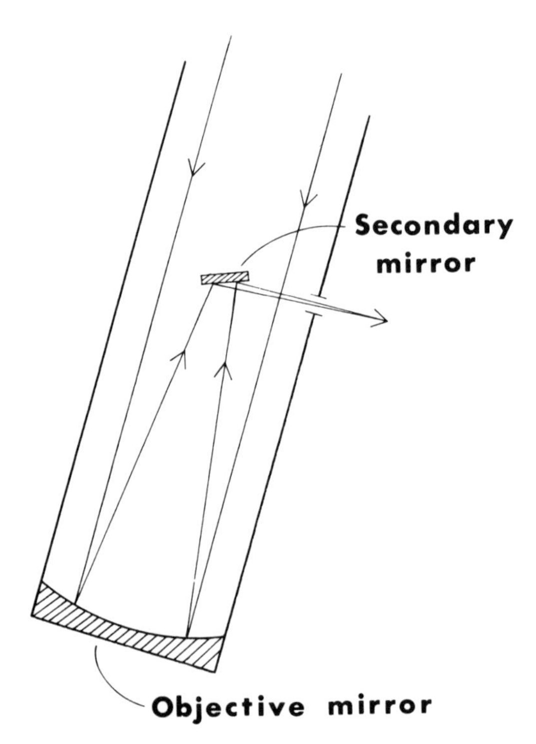

4–7 A reflecting telescope, Newtonian style.
(Andrea K. Dobson-Hockey)

An advantage of the reflector is that it is easier to make. Light does not have to pass through a medium (lens) that is subject to impurities and that absorbs some fraction of the light. However, image quality is more sensitive to irregularities in the objective surface.

Reflectors are better for rendering color because of the **chromatic aberration** of refractors. The focal point that a ray of light is brought to by a lens is a function of the ray's wavelength—that is, of its color. When the red and blue portions of an image are brought to focal points one behind the other, chromatic aberration occurs. Compound lenses, in which the effects of chromatic aberration in the two components cancel each other out at certain wavelengths, are of some use, but planetary observers interested in determining color (such as that of the features on the Moon) almost always use a reflector.

Ultimately, from an engineering point of view, large mirrors are easier to make and to move than are large lenses. So for both astronomical and practical reasons, most of the world's major telescopes today are reflectors. The objective of a reflector must always be partially blocked by a flat mirror **(secondary)** or eyepiece. However, in the largest reflectors an observer can sit at the focal plane without blocking more than a small fraction of the light gathered by the mirror.

MOON MAPS

Although there are other claimants to the title of the first person to look at the moon through a telescope besides Galileo, he was certainly the first to *do* anything with his observations. His estimates of the heights of lunar mountains were reasonable, and he drew what he saw. Although these drawings were rather crude, they were the first documentation of what could be seen and certainly entitle Galileo to priority in the field.

Galileo's drawings did not constitute cartography. The first telescopic Moon map was drawn by Thomas Harriot (1560–1621), a contemporary of Galileo and tutor to Sir

Walter Raleigh. (This map was not published until the twentieth century.) A more impressive work was done by the Belgian mathematician Michel van Langren (Langrenus, 1600–1675). It appeared as *Selenographia* in 1645. The map was a third of a meter in diameter and contained 270 features.

Johannes Hewelcke (Hevelius, 1611–1687) of Danzig created his *Selenographia* in 1647. He used the names of contemporary cities and countries for craters and called mountain ranges after their terrestrial counterparts. This system did not survive intact, but we still have the lunar Alps, Apennines, and Pyrenees today. The mountain ranges on the Moon were named before those of the western hemisphere of Earth were explored. Consequently, we do not see landforms on the Moon with names such as the "Lunar Rocky Mountains."

Hevelius first depicted the lunar **rays,** bright streaks radiating out from some of the large craters. After his death, Hevelius' wife, Elizabeth, continued his astronomical work.

Eutacho Divini (1610–1685) drew a Moon map of minor note in 1649. Two years later another Italian, Father Giovanni Riccioli (1598–1671), published a lunar chart made by his student, Francesco Grimaldi (1618–1663). The map appeared in *Almagestum Novum* in 1651. Riccioli also gave names to the features. He used the names of persons for the craters and assigned them names of famous scientists (mostly astronomers) and philosophers (the distinction between the two not being as great then as it is now) in some way associated with the Moon. The names appear in chronological order north to south and are grouped by nationality, specialty, and so on. There was certainly some arbitrariness to Riccioli's system. He played favorites, giving Tycho, whom he admired, one of the most impressive craters, whereas Galileo, the heretic, received only an obscure, nondescript one. He no doubt named the prominent crater Copernicus begrudgingly, and the two craters Riccioli and Grimaldi are among the largest. Still, despite the fickleness of the process,

Riccioli's system proved popular. Most of the names he chose are still in use.

A German, J. Tobias Mayer (1723–1762), set about to make the first major lunar chart of the eighteenth century. His first goal was to determine the Moon's equator exactly. In trying to do so, he measured the distance from certain features to the eastern and western lunar limbs. He found that these distances appear to change. Galileo himself had said that some features on or near the limb were visible at some times and not at other times, but he went blind before he had time to pursue the matter. Mayer concluded that the Moon appeared to be slowly rocking back and forth slightly so that it is possible to peek around the edges of the visible hemisphere. This phenomenon, called **libration,** allows us to see 59 percent of the Moon's surface (Figure 4-8).

(Lick Observatory Photograph)

4–8 The effects of libration can be seen in these two photographs of the gibbous Moon taken at different times.

Mayer published his work on libration in 1750, but his culminating lunar chart, engraved on copper plates, was not published until 1881, over 100 years after his death.

Johann Schröter (1745–1816) is usually thought of as the father of selenography. Like so many other lunar scientists, Schröter was not an astronomer by vocation. He was a magistrate in the city of Lilienthal. Still, he was able to complete a 10-year study of the surface of the Moon that set his work apart from that which had gone before. Schröter was the first to describe the features he called **rilles.** He also discussed the possibility of a lunar atmosphere.

Drawings Schröter made of selected areas on the Moon appeared in *Seleno-Topographische Fragmente,* completed in 1802. Schröter wanted to document these regions in such detail that future observers could compare them to what they saw on the Moon in order to determine if any changes had occurred. He was, therefore, the first lunar scientist to consider the potential evolution of the lunar surface. Schröter added 70 new names to Riccioli's list of features. Where Riccioli had used one name for a whole group of formations, Schröter devised a system of subsidiary notation to designate individual features.

Wilhelm Beer (1797–1850) was a well-to-do banker in Berlin with an interest in astronomy. He hired J. Heinrich von Mädler (1794–1874) to tutor him in the subject. Soon, von Mädler convinced him to build an observatory, and, although their telescope had only a four-inch aperture, Beer and von Mädler set out to produce a definitive Moon map. They spent 600 nights carefully laying out a grid of lunar reference points and measuring them with a **micrometer.** They then filled in the grid with detailed drawings and measured the heights of various features. Their finished product appeared in 1836 as the *Mappa Selenographia,* which was one meter in diameter. Von Mädler added 140 new feature names to those of Hevelius, Riccioli, and Schröter. Expanding upon Schröter's nomenclature, subsidiary craters, the positions of which had been accurately measured, were denoted by capital Roman letters, the rest of the craters by lower case letters, and peaks by Greek letters.

Beer and von Mädler's work remained a standard reference until 1874 when a British engineer, James Nasmyth (1808–1890), and an astronomer, James Carpenter (1840–1899), teamed up to publish *The Moon: Considered as a Planet, A World, and a Satellite.* The book featured photographs of models they had made of lunar features. Nasmyth and Carpenter summarized lunar knowledge to that date and for the first time delved into the "why" of the appearance of the lunar surface. For this reason they did not remain selenographers but became selenologers.

Meanwhile, Johann Schmidt (1825–1884) was taking up the work of Wilhelm Lohrmann (1786–1840). Lohrmann had intended to produce a true *topographical* map of the satellite but was forced to abandon the project when his eyesight failed. Schmidt produced a map three meters in diameter that contained an astonishing 40,000 individual features. It was published, along with Lohrmann's earlier work, in 1878 by Schmidt with the help of the Prussian government.

While comparing the features on his map to previous charts, Schmidt noticed something odd about the crater Linné. It was gone! Where earlier maps showed a deep crater, we now see only a white spot. Schmidt used this case as an example of change on the Moon, but today it is suspected that Linné was an optical illusion caused by a peculiar effect of the Sun's angle and never deserved to be called a crater in the first place.

By the time of the publication of Walter Goodacre's (1855–1938) *Map of the Moon* in 1910, the face of the Moon was known in far more detail than the face of the Earth. (The interiors of Africa and Antarctica still appeared as blank regions on the globe during this time.) However, by now the link between the eye and the hand had been interrupted and superseded by the invention of photography.

Louis Daguerre (1789–1851) himself took a daguerreotype of the Moon in 1839, but he had to convince people that the image was not just a bright smudge on the plate. William (1789–1858) and George Bond (1825–1865) had better success in 1850 when they produced a recognizable picture. A true photograph followed in 1857.

The problem with photography was that the Moon was a relatively faint object to take a picture of even through a telescope. What's more, the image had to be made big so that the coarse grain of the plate emulsion would not interfere with surface detail. This meant that the available light was stretched over a large area. Because the emulsions then in use were fairly insensitive, it took upward of a minute to produce an image. During this time the body moved in the field, or even if the movement was negated by tracking, the image was severely degraded by vibrations, stray light, or atmospheric conditions. In 1871 the invention of the dry bromide photographic plate reduced exposure times to less than a second at the focal plane, and astronomical photography came into its own.

Soon the major observatories of the world were using their large-aperture telescopes to produce high-resolution lunar photographs. The first lunar photographic atlas was put together at the Paris Observatory. It was called *Atlas Photographique de la Lune* (1896). By the turn of the century, Lick, Yerkes, and Mount Wilson observatories were photographing the Moon, and in 1903 William Pickering (1858–1938) of the Harvard College Observatory published a *Photographic Atlas of the Moon,* which contained images of the Moon made at different phases and, hence, different angles of illumination.

As early as 1921 a problem in lunar nomenclature had become apparent. Some lunar features had as many as three names, depending on how many maps one consulted! The International Astronomical Union (IAU) convened a committee to look into the problem. The result was *Named Lunar Formations* published in 1935. This work was also called the *IAU Map.* The standard convention of the map followed von Mädler as much as possible. Still, there were flaws in the system. The University of Arizona's Lunar and Planetary Laboratory (LPL), under the direction of Gerrard Kuiper (1905–1973), undertook improving and extending the *IAU Map*. This work was finished in 1966. The nomenclature used was adopted officially by the IAU in 1967.

As a postscript to the long succession of telescopic selenography that began with Harriot, Langrenus, and Hevelius in the quest for the "best" Moon map: In the early 1960s the United States Air Force Aeronautical Chart and Information Center undertook mapping the Moon. However, by then events had already begun to surpass the telescopic technique. Spacecraft were soon to view the Moon in a way no Earth-bound observer ever could.

LUNAR FEATURES

During a Full Moon the lunar maria and rays are the most obvious of all the lunar features. This is a time when the Sun's light falls nearly perpendicularly to most of the lunar surface that we see. Shadows are washed out. Features are noticable not because of their relief but because of their intrinsic lightness or darkness.

The maria are really quite dark. They reflect only 6.5 percent of the sunlight that strikes them. That is, their **albedo** is 0.065. (This is the same as that of volcanic obsidian.) Because of this they contrast sharply with the rest of the Moon, the lunar **highlands** or *terrae,* which vary in albedo but are usually between 0.10 and 0.15. The face of the "Man in the Moon" is really composed of four dark maria: Mare Serenitatis (Sea of Serenity) and Mare Imbrium (Sea of Rains) are his eyes, Mare Nubium (Sea of Clouds) is his mouth, and Mare Frigoris (Cold Sea) is his furrowed brow.

There are actually 30 such maria—most of them in the northern hemisphere. These eternally dry seas cover almost half of the lunar nearside. They are usually roundish. Although the telescope reveals that there are craters in the maria, they are much less dense there than in the adjacent highlands. The maria lack other topographical features, for the most part, and are essentially as they appear: large, flat expanses.

Traditionally, maria in the western hemisphere (that part of it which is visible to us) have names that represent moisture: for example, Mare Vaporum (Sea of Vapors) and Mare Humorum (Sea of Moisture). In the eastern hemisphere, the names reflect the nature of a calm sea: for example, Mare Tranquillitatis (Sea of Tranquility) and Mare Serenitatis (Sea of Serenity). An exception is the Mare Crisium (Sea of Crisis). Offshoots of the major maria are called **sinii** (bays). Thus, we have the Sinus Iridum (Bay of Rainbows) and the Sinus Roris (Bay of Dew) just off the Mare Imbrium in the west. We also see the Lacus Mortis (Lake of Death) and Palus Putridinis (Marsh of Decay) on the Moon.

The lunar rays are much lighter than their surroundings. They are almost entirely an albedo feature in that they cast no visible shadows at any time. Furthermore, they are indifferent to topography. They radiate out from craters and drape over maria, mountains, and valleys without deviation. Rays can be 10 kilometers wide and hundreds of kilometers long. The most complete ray systems can be found associated with the craters Copernicus, Tycho, and Kepler (Figure 4–9).

When the Moon is in the first or third quarter, its face is completely different. It is no longer dominated by the maria and rays. Now the shadows take over, making the topographical features stand out in sharp relief. These harsh black shadows contrast against their brightly lit surroundings. They are not like Earth shadows; the lack of a light-scattering atmosphere makes them sharp and total. The lunar shadows actually sharpen the relief of the Moon when it is viewed through a telescope.

Lunar mountains are impressive. They are most obvious when they are near the terminator, and the shadows that they cast are many times longer than their heights. Lunar mountains also affect a lunar eclipse. As totality approaches, there is often a moment, just before the disk of

(Photograph courtesy of New Mexico State University)

4–9 The Full Moon.

the Moon covers the Sun, when sunlight shines between the mountain peaks that happen to be situated on the lunar limb. This effect produces a partial ring of beads of sunlight. The phenomenon is called **Baily's Beads** after the man who first documented it. It serves as a proof of the rough ("imperfect") nature of the Moon.

Lunar mountains are comparable in size to those on the Earth. However, they appear more dramatic. There are several reasons for this. First, we rarely have the opportunity to look at Earth mountains from a vantage point above them. More important, Earth mountains do not always reveal their true height. For instance, most of the mountain that is the island of Hawaii is under water. No such problem exists on the Moon. Also, the Moon is a smaller body than the Earth, so a mountain on the Moon appears to stick up farther than one of similar height on the Earth. On Earth, mountains are frequently surrounded by other mountains or a slightly lower piedmont region. On the Moon they often just jut up out of the lunar plain. The major reason, though, is that the lunar mountains simply appear more striking. They are

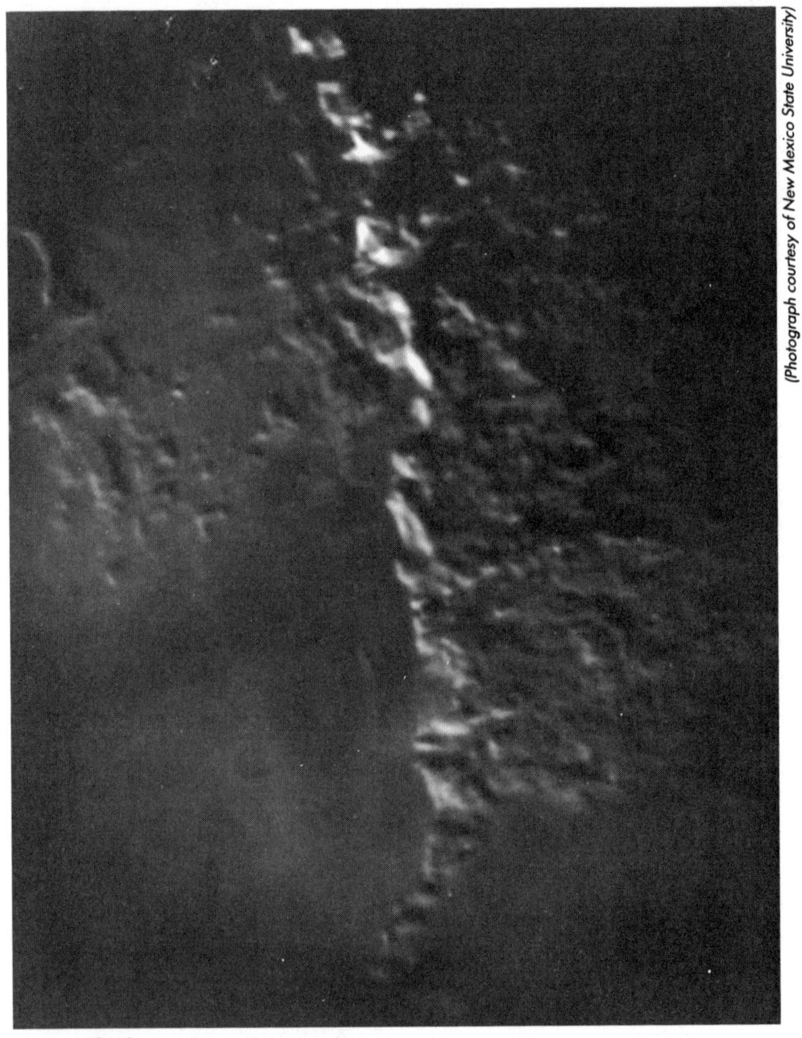

(Photograph courtesy of New Mexico State University)

4–10 The lunar Apennines.

"sharper." They have not been rounded down very much by erosion, as Earth mountains have, or sagged by Earth's much greater gravity. We see them as they were born, jaggedly stabbing at the sky. (Figure 4-10).

More than 30,000 craters are visible from the Earth, all of which give away their relief by their shadows. Craters are unlike anything we usually see on the Earth. They are circular depressions with irregular rims surrounding them. These rims, which are the crater walls, gently slope upward outside of the crater, come to a peak, and then fall off steeply inside. In some lunar craters, the inner walls are terraced.

Craters range in size from tens of kilometers in diameter to small depressions in the ground that you could trip over—and even smaller. Some large craters have central peaks, mountains rising out of the middle of the crater floor. Sometimes there is a cluster of center peaks, as in the crater Tycho (Figure 4-11). At no time do these central peaks rise above the crater rim.

Craters appear randomly over the lunar surface although they are less dense in the maria. Sometimes craters overlap, and the edge of one will intersect or occlude that of another. Craters may also be nested; smaller craters can be found in the floors of bigger ones, and so on.

There are many more small craters than large ones. Still, 150 craters are visible to us that have diameters greater than 80 kilometers. (Eleven are more than twice this size.) Craters 100 miles or so in diameter have historically been called **walled plains.** They are often bounded by a system of multiple concentric walls.

Besides craters, there are other depressions on the Moon. Large crevasses called rilles (or clefts) crisscross the Moon (Figure 4–12). More than 1,000 rilles have been mapped. They are a few kilometers in width and may be up to several hundred kilometers long. Some rilles are quite straight and cross mountains and valleys; sometimes they intersect with each other. When they are gently curved, these rilles are called **arcuate** rilles. Another kind of rille meanders

(Photograph courtesy of New Mexico State University)

4–11 The crater Tycho.

and is called a **sinuous** rille. These rilles are also called "serpentine rilles" because they often begin in a head-like crater and then taper away from it.

Scarps are cliffs on the Moon (Figure 4–13). A rille will often end in a scarp. There are various kinds of ridges on the Moon. One, the **wrinkle ridge,** also called a mare ridge, follows a concentric pattern around some maria.

Now that we are familiar with the Moon's geographical formations (craters, highlands, maria, mountains, rays,

4–12 The Ariadaeus Rille southwest of Mare Serenitatis. The deep crater at center right is Julius Caesar.

rilles, and scarps) (Figure 4–14), let's take a tour of the lunar Nearside, starting in the west. We begin at the far limb where we see part of a mare peeking around the edge of the disk. This mare, named Orientale, is surrounded by three rings of mountains. This paradoxical name comes from the fact that this mare faces east as one views the Moon in the sky and is a holdover from when the Moon was thought to be merely a part of the sky rather than a place with cardinal points of its own.

(Lick Observatory Photograph)

4–13 The Altai Scarp southwest of Mare Nectaris.

Nearby are the nearly twin, walled plains of Riccioli and Grimaldi. The walls of Grimaldi are uneven and rise between 1,000 and 3,000 meters above their surroundings. Grimaldi is 240 kilometers in diameter and is one of a few craters that can be seen by Earthshine even when the Moon is new.

North of Grimaldi, Hevelius and Lohrmann are immortalized. Hevelius borders the Oceanus Procellarum (Ocean of Storms), the largest of the lunar maria, which

(Photograph courtesy of New Mexico State University)

4–14 Some lunar features.

covers over 6 million square kilometers. It is dotted with familiar names: the craters Galilei, Marius, and Mayer. In the north are Aristarchus and Schröter's Valley. The latter is 4 to 5 kilometers wide and 1 to 2 kilometers deep. In the middle of its winds a sinuous rille.

On the eastern "shore" of Oceanus Procellarum is one of the most spectacular lunar features, the crater Copernicus (Figure 4–15). It is located near the center of the disk and is sometimes visible to the naked eye. Copernicus is 93 kilometers in outside diameter, and its floor is 70 kilometers in diameter. Its interior sides consist of a series of **slump terraces** where rock has slipped away. Its walls are more than 4,000 meters high, and the tallest of its cluster of central

(Photograph courtesy of New Mexico State University)

4–15 The crater Copernicus.

peaks tops out at 300 meters. Copernicus' massive ray system actually overlaps that of Kepler, a crater 32 kilometers in diameter that is in the center of the Oceanus Procellarum.

The Oceanus Procellarum appears to drain into two smaller maria "below" or south. These are Mare Humorum and Mare Nubium. Mare Nubium has a unique feature called the Straight Wall. This long, even scarp is 200 meters high and has a face set at approximately 40 degrees from the perpendicular all along its 130-kilometer span.

Following the central meridian of the Moon south is a remarkable string of craters that begins with the walled plain Ptolemaeus, which is 150 kilometers in diameter. (See Figure 4–16.) Sharing a common wall with Ptolemaeus is Alphonsus (of which more will be told later) to the south with a diameter of 120 kilometers. Next in the chain is Arzachel, which is 100 kilometers across. These craters form a lesson in assigning relative ages to craters. Arzachel has the sharpest sloped walls (they are between 5,000 and 3,000 meters high) and has been subject to the least erosion. It is probably the youngest. Alphonsus is older and shows evidence of lava flooding (discussed in Chapter 5). Finally, Ptolemaeus is the least well defined and may be the oldest of the three.

Ptolemaeus is young, though, compared to the crater Regiomontanus. This feature, south of Arzachel, is almost indistinguishable from its surroundings and is obscured by newer craters that are superimposed over part of

(Photograph courtesy of New Mexico State University)

4–16 Ptolemaeus, Alphonsus, and Arzachel.

it. Hipparchus is a similarly dilapidated crater east of Ptolemaeus.

Traveling into the eastern hemisphere, we come to three closely grouped maria: Mare Tranquillitatis (Sea of Tranquillity), Mare Nectaris (Sea of Nectar), and Mare Fecunditatis (Sea of Fertility). Langrenus gets his due with a crater just to the east.

North of Mare Fecunditatis is the nearly perfectly circular Mare Crisium (Figure 4–17). This feature is just

(Photograph courtesy of New Mexico State University)

4–17 Mare Crisium stands out clearly in this picture of the waxing crescent Moon.

visible without a telescope and was named the Gulf of Hecate by Plutarch.

The southern part of the lunar Nearside is dominated by the crater Tycho. Its ray system, the largest on the Moon, extends 3,000 kilometers from Mare Nectaris in the east to the Oceanus Procellarum in the west. Tycho is 85 kilometers in diameter, and its walls stand 4,000 meters high (the central peaks reach 1,500 meters). It is believed that Tycho is one of the youngest craters on the Moon.

Due south of Tycho is Clavius (Figure 4–18). This immense walled plain is 225 kilometers in diameter and spans 5 kilometers in elevation from rim to bottom. Clavius is so big that it has a respectable crater inside it: Rutherfurd, which is 50 kilometers across.

(Photograph courtesy of New Mexico State University)

4–18 The crater Clavius.

Just to the north of Tycho is Hell (named after Father Maximilian Hell).

The largest walled plain is Baily near the South Pole. It is 294 kilometers in diameter and 5 kilometers deep, but it does not look as impressive as Clavius since we see it projected at a shallow angle.

In the northern latitudes of the lunar Nearside we see two principal maria, Mare Imbrium and Mare Serenitatis, separated by the lunar Apennine and Caucasus mountains. A gap between these two ranges connects the two seas. The Apennines are 700 kilometers in length and are as high as 6,000 meters. They are divided by several deep valleys.

Mare Imbrium is the site of several major isolated mountain peaks. A line connecting two of these, Pico and Piton, and the Straight Range follows the outline of the border of Mare Imbrium. A curious little mare called Sinus Iridum joins Mare Imbrium and is two-thirds surrounded by a single mountain range, the Juras.

The northern rim of Mare Imbrium is defined by the lunar Alps. A huge gorge called the Alpine Valley cuts through them radially to Mare Imbrium and connects that sea with Mare Frigoris. This fissure is the largest of its kind of the lunar Nearside.

The crater Plato, 100 kilometers in diameter, is superimposed over the lunar Alps. Its extremely dark floor contrasts with its lighter, rough surroundings.

The north polar region is the home of craters named for Greek astronomers such as Thales, Aristotle, Eudoxus, Pythagoras, Meton, and Anaxagoras.

This ends our brief tour of the lunar Nearside. Of course, there is another side to the Moon, the lunar Farside, that we cannot see with Earth-based telescopes but that has an equally varied geography. It remained a complete mystery to astronomers and cartographers until spacecraft photographed it in the 1960s. Not until then did Moon maps appear that had two disks on them instead of one.

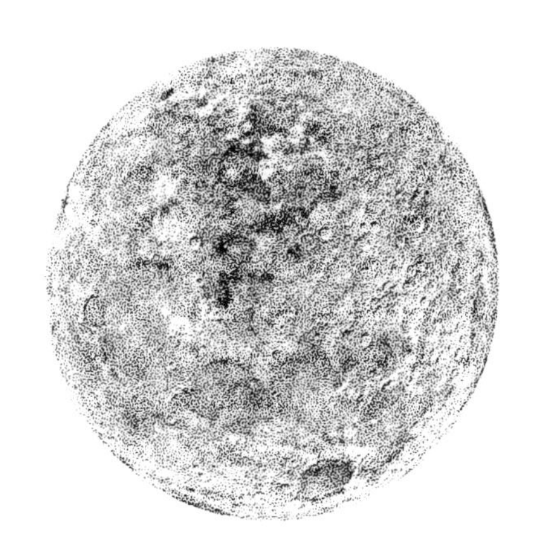

5
THE MOON FROM A DISTANCE

GRAVITY AND TIDES

While the Moon was steadily being mapped during the seventeenth and eighteenth centuries, its orbit was also being charted. The details of its motion had to remain empirical, though, until a single question was answered: If the Moon is orbiting around the Earth, what keeps it there and why does it not fall down? The question could just as well be: Why does the Moon not go flying off on its own somewhere? What keeps it here? Copernicus, Brahe, and Kepler had done away with crystalline spheres, but something was needed to take their place.

The answer lay in Sir Isaac Newton's (1642–1727) invention (or discovery, if you prefer) of universal **gravitation.** Newton described gravity as the attraction that all massive (used in this sense as opposed to "massless") objects have for each other. The apple is attracted to the ground. The ground is attracted to the apple. The apple succumbs to this attraction more readily because the Earth is so much more massive than the apple. We see the apple fall; the Earth's tiny motion toward the apple is imperceptible. Newton was able to state that the attractive force of a body is directly proportional to its mass. He maintained that it is inversely proportional to the square of the distance to the object being attracted.

Newton's gravity was but one of a kind of phenomenon called *forces.* **Newton's Laws of Motion** state that an object in motion tends to remain in motion (traveling at the same rate and in the same direction) until it is acted upon by a force (gravity, a magnet, a good swift kick, etc.) The same holds true for an object at rest. It will stay that way until it is acted upon by a force.

Newton saw that the Earth's gravitational force must extend to the Moon and vice versa. The Moon is traveling at a certain speed. It is also attracted toward the Earth. The result of this tug of war is that the Moon remains on its path, but the path is curved into a closed ellipse.

This is somewhat unsettling. It may still seem as though the Moon should fall out of the sky someday. Let's try an illustration to clear it up. Instead of the Moon, suppose we have a slingshot and a stone. If we fire stone into the air, the stone will obey Newton's Laws. That is, it will try to keep going at the velocity given it initially by the slingshot. (In this sense, "velocity" is both a speed and a direction). All the time, though, it is being acted on by gravity. The stone is attracted by gravity to the Earth. The stone slows and finally falls back to the ground. Now suppose that the stone is shot at an angle. It travels upward, but it also travels horizontally (perhaps toward a target). The vertical motion of the stone is affected by gravity as before; the center of the stone is attracted toward the center of the Earth. However, the motion of the stone perpendicular to the line connecting the center of the stone to the center of the Earth is not bothered by gravity. The motion in this direction continues unhindered until the stone is stopped by striking the ground. The faster and higher the stone goes, the longer it takes for gravity to pull it back to Earth and the longer it has to travel unimpeded in the horizontal direction. The trajectory of the stone is a combination of vertical and horizontal motion.

On a flat and infinite Earth the stone would always hit the ground, no matter how hard it was thrown. The Earth is not really flat, though. Imagine now, a giant "slingshotter" standing on the Earth. The giant is so tall that she can see the curvature of the Earth falling away in all directions. With her sling she shoots a stone the size of the Moon. Remember that the stone will always be attracted toward the center of the Earth. Because the Earth is round, this direction will keep changing.

Now we will introduce a tool for aiding understanding. Sometimes it is helpful to think of something that happens continuously as happening in discrete steps. Once we can understand how the steps work, we can easily imagine the steps to be so small that they appear to be continuous again. With this in mind, we return to the giant and the flying stone. The stone begins by moving horizontally. Initially, this direction is parallel to the surface of the Earth. As the stone moves to one side, though, this is no longer the case. The stone continues to be pulled toward the center of the Earth, but that center is no longer perpendicular to the stone's horizontal path. The effect of the Earth's gravitational tug is to bend the path toward the Earth. Once this has happened, the stone is again traveling in a direction perpendicular to the line connecting it and the center of the Earth. An instant later, though, it has moved again. Once more, gravity changes its direction. Gravity continues to bend the path of the stone downward, but as it does, the Earth's surface curves away under it. Now if the giant has shot the stone in just the right way so that the path taken by the stone is bent into a curve that just matches the curve of the Earth, we can see that the stone will travel all the way around the world. It will always be pulled toward the center of the Earth by gravity, but the Earth will always curve away under it by just enough to make up for the distance the stone falls. When it gets back to its starting place, it will be no closer to the Earth than it was to begin with. The result is obvious (and perhaps painful).

What's more, unless it hits the giant on the head, the stone will continue to orbit around the Earth. It can do nothing else because there is nothing to stop its forward motion (always perpendicular, remember, to the line connecting stone and Earth and therefore impervious to their mutual gravitional attraction). Once again, an object in motion tends to remain in motion. . . .

The stone (Moon) we have described is in a circular orbit around the Earth. The giant could have slung it in a slightly different combination of directions and speeds (horizontal and vertical) to obtain an elliptical orbit. The idea is the same.

Except for how it got there, the stone is the Moon. This stone will continue in the same orbit until something or someone applies a force to it. If the radius of its orbit increases somehow, it will slow down (Kepler, again), and a new equilibrium will be reached. There *is* another force acting on the Moon that will indeed move it from its orbit, but to tell this story we must first go back to ancient times.

We have seen before that a link between the Moon and the tides was apparent to all but the most primitive of seacoast dwellers. A high tide (one of two each day) occurred every 24 hours, 50 minutes. This was also the time between **transits** of the Moon. Exactly what the nature of the connection was between these two phenomena, though, was less apparent. The Greek explorer, Pytheas (fl. 300 B.C.), said that the Moon caused tides, but he did not explain how. It was the Marquis de Laplace (1749–1827) who applied Newton's new theory of gravity to the tides and found that with it he could explain them very nicely.

Consider three things: the Earth, an ocean on the side of the Earth closest to the Moon, and an ocean on the opposite side. Because they are liquid, the oceans flow. The near ocean is closer to the Moon than either the center of the Earth or the far ocean. Because of the inverse square dependence of gravity, a stronger force is tugging on this ocean that is closer to the Moon than the force that is pulling on the Earth. This difference in force causes the near ocean to be pulled away from the Earth. Meanwhile, the gravitational force from the Moon is exerting a pull on the far ocean *weaker* than the force being exerted on the Earth. This time the force differential pulls the Earth away from the ocean.

The result in each case is the same: The middles of both oceans appear to be pulled out away from the Earth. At any given time there exists such a **tidal bulge** on opposite sides of the Earth. The ocean is actually slightly deeper in the bulge. Where does the extra water come from? Also at any given time there are places on the Earth (approximately 90° to either side of the line connecting the Earth and the Moon) where the ocean and the center of the Earth are equidistant. There is no gravity differential here, but because water flows and all the oceans are connected, these parts of the oceans lose water to accommodate the tidal bulges. In these parts of the ocean, the water is slightly shallower than the mean and a tidal trough is present.

Remember that the Earth is rotating and does so much faster than the Moon is orbiting the Earth. The tidal bulge remains on a line connecting the centers of the Earth and the Moon. Therefore, as the Earth turns, different parts of its surface experience the tidal bulge. From our point of view on the Earth, it seems as if the tidal bulges are moving around the Earth once every day or so. Say, now, that we are on a particular island or seashore that is rotating with the Earth into the tidal bulge. The water gets deeper and comes up higher onto the land. We say that the tide has come in. As our parcel of land comes out of the tidal bulge and into the trough, the water level goes down. The tide goes out. We now encounter the second tidal bulge and the second trough. The same sequence happens, and the cycle begins again. Thus, in a little more than a day, every place experiences two high tides and two low tides. If you happen to be at the end of a narrow bay near where the water of the incoming tide piles up, the effect is magnified (Figure 5–1).

The Sun causes tides, too, a possibility mentioned by Pliny. It does so in just the same manner as the attraction between the center of the Earth and the center of the Moon. The Sun is much more massive than the Moon, but it is so much farther away that its gravitational force, and hence the gravitational differential it causes, is less than that of the Moon. Still, the net tide we experience is the sum of the

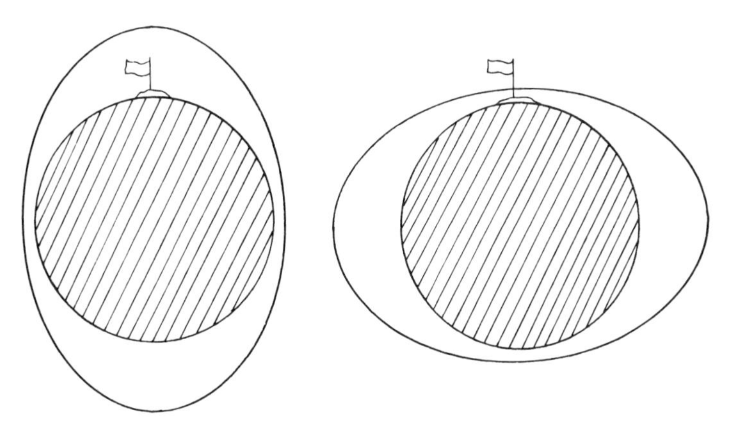

5–1 High *(above)* and low *(below)* tides are shown for an imaginary (and very wet!) island.
(Andrea K. Dobson-Hockey)

lunar and solar tides. When the Moon and the Sun are both on the same side of the Earth (during the New Moon), their effects add, and we have a high tide that is higher and a low tide that is lower than usual. This is a **spring tide.** When the Moon and Sun are on opposite sides of the Earth (during the Full Moon), their tides cancel each other out to some extent, and we experience a tidal amplitude that is less than usual. This is the **neap tide.**

The Moon causes a tidal bulge in the solid Earth as well, albeit one of only 20 centimeters (compared to 60 centimeters in deep ocean). The difference occurs because the tensile strength of rock is so much greater than that of water. Because the solid Earth does not flow as well as the oceans, this tidal bulge cannot slip around the Earth as well either. As the Earth rotates, it takes the bulge with it. It takes time to stretch and bend the Earth. It takes time for the new bulge (on the Earth–Moon line) to form and the old bulge (now rotated off the Earth–Moon line) to dissipate. This introduces a time lag (Figure 5–2). The tidal bulge of the solid Earth is always ahead of where it ought to be—that is, it is not directly below the Moon. The bulge acts as a lever arm on the Moon. Gravitational attraction between the Moon and the bulge is no longer acting in the same direction as the attraction between the center of the Earth and the center of

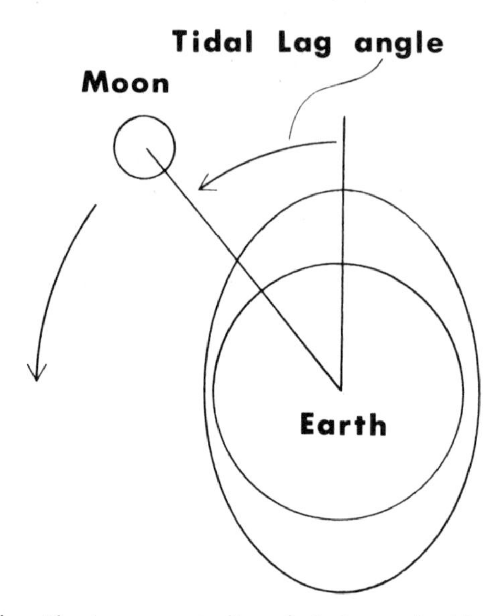

5–2 Tidal lag. The (exaggerated) angle between the Moon-Earth line and the line connecting the two tidal bulges is caused by the time it takes to form the bulges. New bulges are being raised directly beneath the Moon, but the satellite will have moved through the tidal lag angle again before they reach their maximum height.
(Andrea K. Dobson-Hockey)

the Moon. Nevertheless, the two are drawn to each other, and the Moon tends to be pulled forward (toward the bulge). At the same time, the Moon tugs on the bulge. This pull opposes the direction in which the Earth rotates. The Moon applies a torque to the rotating Earth, the effect of which is to slow the Earth down! This effect is miniscule, but over millions of years it adds up. The day is now more than 1 second longer than it was 100,000 years ago.

To examine what happens next, we need to take a look at one of the guiding laws of the universe, Conservation of Momentum. It turns out the produce of a body's mass and its velocity, called its **momentum,** must be conserved. This conservation extends to a system of objects as well. If two bodies interact so that one of them loses momentum, the other must gain it. An example is a traffic accident in which a moving car "rear ends" a parked one. The first car has momentum (it has both mass and velocity), but the second does not (it has no velocity). At impact, the first car imparts momentum to the second car. The first car slows, and as a result the second car must start moving to conserve momentum. This is indeed what happens.

Our example referred to bodies (cars) moving in straight lines. There is the law of Conservation of Angular Momentum as well. A body's moment of inertia about some axis is a function of its mass and the square of the distance from the axis:

$$I = M \times r^2$$

for a body revolving about a point, and is

$$I = \frac{2}{5}M \times r^2$$

for a spherical body revolving about its center. The product of the angular velocity and the moment of inertia of an object or system of objects must also be conserved in the absence of outside forces. The Moon and Earth are such a system. Both have angular momentum because they rotate. The Moon has additional angular momentum because it revolves around the Earth. When the Earth slows because of the Moon's torque on its tidal bulge, it loses angular momentum. This angular momentum must go elsewhere in the system. It can't just disappear! It is the Moon that takes up the angular momentum. It does this by increasing the radius at which it orbits the Earth. The result is that as the Earth slows, the Moon moves away from us. It is currently doing so at a rate of four centimeters per year.

What will eventually happen to the Moon? It was Sir George Darwin (1845–1912) who first examined the evolution of the Moon's orbit (in *The Tides and Kindred Phenomena in the Solar System,* 1898). Darwin was the son of Charles Darwin, who is remembered for discussing evolution of another sort. The junior Darwin concluded that the Moon will continue to move away from the Earth as the Earth's rotation period lengthens. Because it is **tidally locked** to the

Earth (keeps the same face pointed toward it), the Moon's rotation period will slow as well. This process will continue until the Earth day and the Moon day (the month) are both the same, equaling about 47 of our current days. This will happen in approximately 50,000,000,000 years! From this point on, the Earth and Moon will no longer affect each other tidally. In fact, there will no longer be tides as we know them at all. The Earth will keep the same face pointed toward the Moon—just as the Moon does toward us now—and there will be a permanent tidal bulge located under the Moon. The Sun, always a minor player in our story up to this point, will come into the game now. Its gravitational force will cause the Moon to spiral in toward the Earth. Eventually, when the Moon is less than 10,000 kilometers away from the Earth, the differential of the Earth's gravitational force acting on different parts of the Moon will be great enough to tear it apart. Once it has passed this limit, called the **Roche limit,** the Moon can no longer be a solid body. It will disintegrate into small pieces spreading out in a ring around the Earth similar to the rings of the planet Saturn. These pieces will eventually fall to Earth as meteors, and the Moon will be gone forever. (Many billions of years sooner than this, changes in the Sun will radically alter the solar system as we now know it and make the above a scenario of theoretical importance only.)

THE LUNAR THEORY

Although the Moon is not destined to be with us for eternity, it will remain in an orbit much as it is now for a long time to come. For all practical purposes its orbit is a model of regular motion.

In 1474 Johannes Müller (Regiomontanus, 1436–1476) suggested a practical application of the motion of the Moon. During the fifteenth century the first of the great ships were setting out to explore the world. Where they might go or what they might discover, though, was limited, not by the speed of these vessels but by the ability to navigate to where they were going. Regiomontanus wanted to use the Moon as a device for calculating longitude. His plan was to chart its orbit well enough so that one could predict where it would be in the sky against the background of stars at a given time for a given place. Using this information and comparing it to actual observations of the Moon made aboard ship, a mariner could fix longitude.

Regiomontanus' idea was not practical in the late 1400s, buy by the 1600s, when King Charles II of England heard a similar idea, Newton was at work making orbital prediction a mathematical science. Depicting the Moon's motion became fair game. The king appointed the Reverend John Flamsteed (1646–1719) (Figure 5–3) to a committee to study the plan. The committee concluded that what was needed first was a good catalogue of star positions against which to fix the Moon. The Royal Greenwich Observatory (Figure 5–4) was founded for this purpose. Flamsteed, as the first Astronomer Royal, worked on mapping the stars. Sir Edmond Halley (1656–1742) (Figure 5–5), his successor, worked on the second problem, the motion of the Moon.

(Photograph courtesy of the Royal Greenwich Observatory)

5–3 John Flamsteed.

Halley looked at ancient eclipse records made by Hipparchus and Arabic scientists and found that the period of the Moon seemed to be speeding up. This came to be called the **secular acceleration.** Richard Dunthorne (1711–1775) calculated that it was equal to 10 seconds of arc per century. Clearly, the secular acceleration had to be taken into account in predicting lunar positions accurately, even though its cause was a mystery.

Measuring the Moon's position and extrapolating into the future would always remain inaccurate simply because the technique of measurement itself could never be infinitely precise. One could never fully trust such knowledge gained empirically. The ultimate table of lunar positions had to be based on mathematical calculations derived from Newton's theory. Observations would then be made only to confirm the calculations. In this way the Moon's orbit became a problem in mathematics rather than in astronomy, and the quest for a successful **lunar theory** attracted some of

(Photograph courtesy of the Royal Greenwich Observatory)

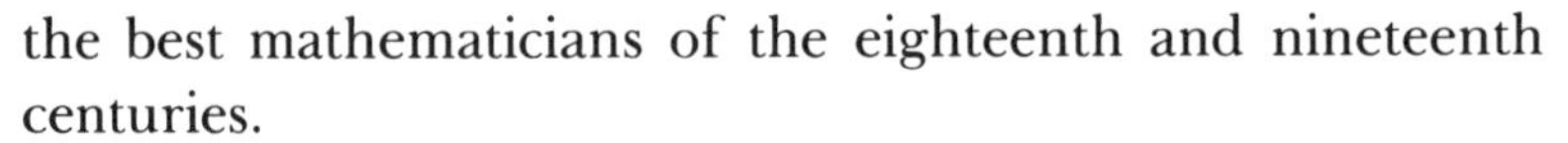

5–4 The Octagonal Room, Royal Greenwich Observatory circa the 1600s.

(Photograph courtesy of the Royal Greenwich Observatory)

5–5 Edmond Halley.

the best mathematicians of the eighteenth and nineteenth centuries.

Ironically, the Moon was too open to inspection. Every tiny effect and perturbation of the Moon's orbit could be easily observed and therefore had to be taken into account by the theory. Thus, although Newton himself had had great success in calculating the orbits of the Jovian and Saturnian satellites, the Moon would not behave. Even when the known inequalities (evection, variation, and the annual inequality) were incorporated into the theory, the Moon failed to move at the rate Newton calculated that it should, and he gave up, discouraged.

Others tackled the problem after Newton and added effects thought to account for the secular acceleration. These were all treated properly according to gravitational theory and were often defined as an added term in the master equation for the motion of the Moon.

Alexis Clairaut (1713–1765) and Jean D'Alembert (1717–1783) of France actually published a table of lunar positions based on theory alone. Clairaut became the first to calculate the proper motion of the Moon's perigee (the thing that stumped Newton) in this way: He was so confident in his calculations that when his results disagreed with observations, he said that it was Newton's theory, not Clairaut's mathematics, that was at fault. Clairaut suggested that the gravitational equation was slightly different at close range. He thought that there should be an extra term inversely proportional to the *cube* of the radius. Finally, Clairaut discovered an error in his figures that brought the lunar perigee neatly within the realm of gravitation. This was a major coup for Newton's theory.

(Photograph courtesy of the AIP Niels Bohr Library)

5–6 The Marquis de Laplace.

The problem of the secular acceleration was still a nagging one. The Marquis de Laplace (Figure 5–6), a French mathematician, solved at least part of the mystery. After puzzling unsuccessfully with a red-herring noninstantaneous action of gravity, he dropped exotic effects and concentrated on Newtonian physics again. Laplace found that it is the gravitational effect of the planets that is to blame. They are reducing the eccentricity of the Earth's orbit around the Sun. This in turn increases the Sun's perturbation on the Moon. Laplace's work was summarized in *Mecanique Celeste* (1802).

Laplace was able to compile a set of lunar tables that were accurate to within a minute of arc. However, by this time that patient observer, Tobias Mayer (after mapping the Moon), had come up with a semiempirical formula with only 14 terms in it that nailed the accuracy down to a value essentially the same as that of Laplace. Theory was still behind observation.

(Photograph courtesy of the AIP Niels Bohr Library)

5–7 Andreas Hansen.

In 1857, Andreas Hansen (1795–1874) of Gotha (Figure 5–7) finally published a table, based on his calculations and confirmed by a century of observations, that was accurate to within 1 or 2 seconds of arc per century. A modified version of Hansen's theory remained the standard for 50 years. There were still problems, though. After about 1860, the Moon's motion could be detected to deviate from that predicted by Hansen.

George Hill (1838–1914) looked at the lunar equation as the gravitational interaction of three bodies rather than that of two with some perturbations added, as had been done before. Ernest Brown (1866–1938) used Hill's methods for formulating a lunar table in 1919 of exceptional, but not infinite, accuracy. By this time, though, it was known that it was not enough to keep track of only the Moon in developing the lunar theory.

We already know that Darwin prepared the mathematics describing the retardation of the Earth by the Moon. The Earth is losing energy because of the friction caused by the tide dragginng the oceans across the seabed. Shallow bodies of water like the Bering Sea are especially efficient at producing tidal drag. These tidal effects slow the Earth. As the Earth slows, the Moon appears to speed up. This "acceleration" of the Moon had to be added to the lunar theory.

As the lunar theory became more and more complex, it became apparent that a major factor in producing the unreliability of the lunar tables was not the Moon but the Earth. Regrettably, it was the Earth that was used to measure time. The Earth "clock" was not only slow; it was unreliable.

The Earth's rotation changes, and it may do so abruptly. Instead of being a smooth sphere, the Earth is a complicated, amalgamated body with a dynamic atmosphere, shifting faults, and drifting ice fields. Each of these can change the shape of the Earth. This change of shape alters the mass distribution for the calculation of its moment of inertia. Because angular momentum must be conserved, the Earth's period of rotation changes to make up for the refiguring. It does this continuously and irregularly as it responds to various changes on its surface. These changes affect the Earth–Moon system. Thus, a perfect lunar theory of a single equation to express the motion of the Moon is ultimately impossible. There can never be enough terms in it to account for all of the unpredictable wobbles and shudders on the Earth.

The failure of the lunar theory is not the fault of the mathematicians who labored long and hard over it; it is caused by the lack of a stable mount on which to stand to take our measurements. Consider, though, what has been achieved. Even the accuracy of Hansen's 1 second of arc per century is an angular separation equal to one 1800th of the disk of the diameter of the Moon in the sky or that of a pinhead seen at a distance of 200 meters! Today, sophisticated computer programs can pinpoint the Moon's location at a given time to the kilometer—an accuracy sufficient to send men to land in a prescribed place on it and return them safely from there.

There is an irony here: The lunar theory never was commonly used for the accurate determination of longitude, as had been proposed. Alternate methods were always more accurate, and finally the invention of the **chronometer** supplanted it altogether. Modern navigators look at their clocks or listen to radio signals rather than watching the Moon. Today, however, another form of navigator, the celestial navigator, uses for voyages to the Moon itself equations first roughed out by Clairaut and Laplace 200 years ago.

NINETEENTH- AND TWENTIETH-CENTURY LUNAR ASTRONOMY

Although they could completely map its surface and compute its orbit, selenologists wanted to know more about what the Moon was actually like. However, without going there (a possibility that was not even seriously considered until relatively recent times), they were limited to what could be learned remotely. Most of the information came from the only thing that reaches us from the Moon—**electromagnetic radiation,** chiefly in the form of light (Figure 5–8). Imaginative means squeeze out all the information in this radiation.

First, there was the matter of how bright the Moon is. Compared to most astronomical objects, it is very bright indeed. For instance, fuel was conserved in some rural English communities during the last century by enacting ordinances that forbade the lighting of street lights on the two nights a month before Full Moon.

Although the Full Moon is quite sufficient to illuminate our way at night, it is insignificant in the day. We consider the Moon bright but do not realize how bright the

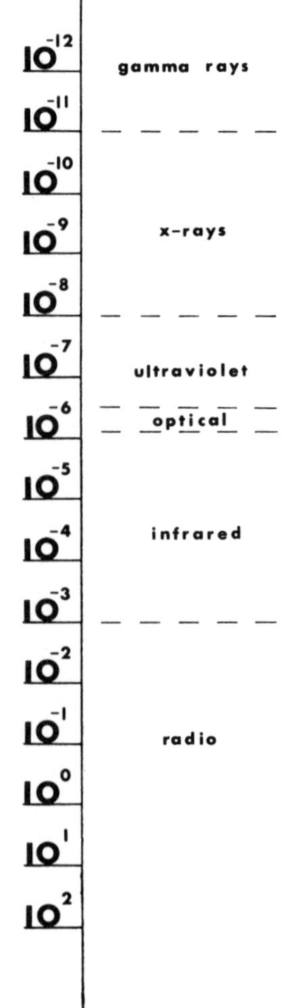

5–8 The electromagnetic spectrum. Of all the wave lengths emitted by radiating objects in space, only those coming through the narrow optical "window" and some radio signals penetrate the Earth's atmosphere. *(Andrea K. Dobson-Hockey)*

Sun is. If the entire sky were filled with Full Moons at night, we would receive only about one-fifth of the light we get during the day from the single Sun. The Sun is actually 456,000 times brighter than the Moon.

Furthermore, the Moon grows darker rapidly as it phases. The Quarter Moon is only about one-tenth as bright as the Full Moon. We might expect it to be half as bright, but the numerous shadows cast by every little surface irregularity cut down drastically on the reflected sunlight in conditions other than direct overhead illumination. The shadows of the mountains and craters that we see are not sufficient to account for this phase effect. The cause is the myriad of tiny shadows that we cannot resolve, right down to those cast by individual grains of sand.

In 1861 J. C. F. Zöllner (1834–1882) attempted to determine quantitatively how bright the Moon is. Zöllner found that although the center of the fully illuminated disk is very bright, the intensity falls off rapidly as one looks closer toward the limb. That is, the intensity of reflected light decreases greatly with a corresponding decrease in the angle at which sunlight strikes the lunar surface. Zöllner calculated the average lunar albedo to be 0.07. The Moon's albedo varies by a factor of 3.5 over its entire surface.

Johannes Wilsing (1856–1943) was one of the first to apply the technique of **spectroscopy** to the Moon. Spectroscopy breaks up light into its component colors and determines intensities at specific wavelengths. Wilsing found, expectedly, that the lunar spectrum is very similar to that of the Sun. The Moon has little color of its own. This may seem strange to those of us who have seen a beautiful yellow or orange Moon rising, but this color is produced by agents in the Earth's own atmosphere. When the Moon is on the horizon, we must look at it through a great deal of atmosphere. When it is high in the sky (and the sky is clear), we see a more realistic whitish color.

In the 1920s Bernard Lyot (1897–1952) studied the **polarization** of light reflected by the Moon. Polarization has to do with the orientation of light's plane of oscillation. Lyot found a polarization similar to that produced by light reflected by volcanic ash.

The fourth Earl of Rosse, Lawrence Parsons (1840–1908), compared the heat coming from the Moon to the energy it received from the Sun and came up with a range of temperatures on the lunar surface. He found that the Moon is an uncomfortable place. The atmosphere of Earth moderates its temperatures, but the Moon, having no atmosphere, is a place of extremes. At the equator daily temperatures may top 110 degrees C., well above the boiling point of water, were any to be found on the Moon. At night the temperatures drop to around −173 degrees C. During a lunar eclipse, the temperature may plunge 150 degrees in an hour.

In his 1638 book, *The Discovery of a World in the Moone,* John Wilkins (later an Anglican bishop) theorized that the Moon might have an atmosphere and weather like the Earth. Unfortunately for Wilkins' thesis, when the Moon occults a star by passing between us and it, the star disappears instantaneously when it is cut off by the lunar limb. There is no dimming effect as would be expected if the star's light was first attenuated by a thin layer of gas at the Moon's edge. Hence, the evidence for a lunar atmosphere was scanty.

The Irish physicist George Stoney (1826–1911) explained why the Moon has no atmosphere and why, if it had one, it would be unable to retain it. Using Newton's Laws, it is possible to determine how high an object will travel when it is thrown upward at a given velocity. Turning this equation around, it is possible to determine the initial velocity necessary to propel an object to a given height. Set this height to infinity, and the resulting velocity is called the **escape velocity.** The escape velocity for a planet is the speed at which an object will never fall back and will leave the planet forever.

The escape velocity for a body is determined by the mass of the body itself. The escape velocity for the Moon is much less than that for the Earth. For this reason, it is harder to launch a rocket from the Earth to the Moon than vice versa.

All molecules of gas, such as those that make up the atmosphere, possess a certain amount of **kinetic energy,** depending on their temperature. Kinetic energy is half the product of the mass of a molecule and the square of its velocity.

$$KE = \frac{1}{2}MV^2$$

Most of the molecules in the Earth's atmosphere have a kinetic energy such that the average molecule travels at a speed less than the escape velocity. This is not the case on the Moon, however. There, most oxygen, nitrogen, and carbon dioxide molecules (the primary constituents of our atmosphere) would travel at an average velocity greater than the escape velocity. They would leap from the Moon and be lost forever. Lighter molecules such as hydrogen, which would travel even faster, would dissipate more rapidly than the heavier molecules. Thus, if eons ago the Moon did possess these gas molecules in quantities great enough to be considered an atmosphere, it would long since have lost them, especially if the lunar temperature were greater then than it is now. Only the very heaviest gas molecules, like the rare gas xenon, could hold out on the Moon for as long as the Moon is believed to have existed.

If the Moon could not have retained any of its original gas constituents, what about a replenished atmosphere? It is possible that some gas trapped within the Moon for millions of years may be released from time to time. Radioactive elements inside the Moon might produce some new gas as well, but an atmosphere sustained by these processes would be very tenuous indeed.

Audouin Dollphus of the Meudon Observatory, France, made a search for twilight effects near the terminator. He used polarization techniques to set a limit on a lunar atmospheric density at 10^{-9} times that of the Earth's. The accepted value today is 10^{-14}, which is a far better vacuum than any attained in a laboratory on Earth.

At this density, the lunar "atmosphere" (for it no longer deserves to be called such outside of quotations marks) must be dominated by transient gases. The Sun is continually emitting a stream of ionized particles in all directions called the **solar wind.** The Earth's magnetic field blocks us from the solar wind, but the unprotected Moon is bombarded by it. Some of these particles may remain bound to the Moon for a while. However, a single **solar flare** would be enough to blow away such a tenuous envelope.

With such a low escape velocity, the Moon has no chance of retaining water. The H_2O molecule is one of only moderate mass. It is unlikely that the Moon ever had much water. Although there are valleys on the Moon that drain into maria and resemble dry river beds and other signs that some material flowed once in the Moon's past, these features can more easily be explained as volcanic lava rather than water. The lunar maria have always been oceans dryer than any desert, and the Sea of Showers has never seen so much as a single raindrop.

Into this hostile lunar environment of hot and cold, vacuum and drought, people have long insisted on placing life—not in reality, but figuratively and with imaginative fervor. The invention of the telescope and its revelation of a lunar surface more Earth-like than had previously been thought bolstered this tendency.

In 1835, the public was so prepared to believe that the Moon was inhabited that the *New York Sun* ran a scandalous article supposedly based on the observations of the prominent astronomer, Sir John Herschel, that described the batmen, birdmen, and other creatures supposedly alive and well on the Moon. The article was so popular that it was reprinted as a pamphlet, and 60,000 copies were sold in the United States before the hoax moved on to a successful run in Europe. Few readers doubted its authenticity. By late in the nineteenth century, hope of highly developed lunar

neighbors had diminished. Still, there were those who were not ready to concede entirely.

If life does exist on the Moon, it has become clear that it must be both hardy and inconspicuous. Today we know that no higher life forms live there. It is interesting to note that if educated people at the turn of the century had been asked to pick a candidate for the first being on the Moon, they would have been much more likely to choose some weird alien creature than the *Homo sapiens* from Wapakoneta, Ohio, who was to make his appearance there just 70 years hence.

WHERE DID THE CRATERS COME FROM?

There is a certain familiarity about the Moon when it is viewed through the telescope. Its features remind us of landscapes on our own world. An exception is the craters. These plentiful features are unlike any depressions on Earth. What are they, and where did they come from?

The only likely Earth analogs for craters are the **caldera** of great volcanoes; thus, there was a tendency to give lunar craters a volcanic explanation. A volcanic origin for the craters implied a geologically active Moon. Examples of such activity can be seen there. Scarps look like **faults,** and in some cases there is a higher-than-normal density of craters near scarps. There are crater chains on the Moon just as there are volcano chains on Earth. There are depressed areas on the Moon that indicate subsidence (or collapse), and the sinuous rilles suggest some sort of plastic flow. In fact, there is evidence of almost every kind of geological activity taking place at some time on the Moon. Did it produce the craters?

Both Schröter and von Mädler accepted the volcanic theory. William Hershel (1733–1822) claimed to have seen a volcano actually erupting on the Moon (in Aristarchus), although it is more likely that he saw the Sun reflecting off of a rock facet on a mountain peak. Such reports abetted the volcanic theory.

The problem was that lunar craters didn't really

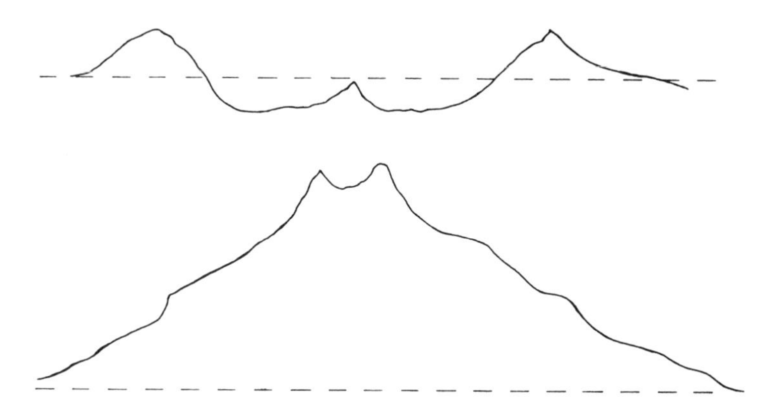

5–9 Profile of a lunar crater *(above)* versus that of a terrestrial volcano *(below)*.
(Andrea K. Dobson-Hockey)

look very much like volcanoes at all. A volcano is dominated by its steeply sloped cone, which features a small crater at the top. This vent is usually the highest point on the volcano. Lunar craters are more flattened than Earth volcanoes (Figure 5–9). They are dominated by the crater depression itself. The floor of the crater is always lower than its surroundings, not higher. Furthermore, even if the crater has a central peak, this peak is always lower than the rim of the crater. Crater basins are relatively flat, not concave like a volcanic vent or caldera. The amount of material in the rim of a crater that rises above its surroundings is usually just enough to fill the crater basin that lies below it. Then there is the problem of size. One of the largest volcanic calderas on Earth is now a lake in Oregon that is 10 kilometers in diameter. Some lunar craters are an order of magnitude larger in size. Why would they be so much bigger?

Nasmyth and Carpenter (see Chapter 4) approached the problem of the morphology of lunar craters in this way: A crater is indeed volcanic but does not exhibit a volcanic cone as on Earth. In the Moon's lower gravitational field, lava spewed out of an opening in the lunar surface like a fountain. The ejecta flew outward and then fell back to the ground at a distance to form the crater rim. Afterward, volcanism might have recurred, this time more conventionally, and formed cones in the middle of the crater floor. These cones are the central peaks.

This theory failed to account for why central peaks are never higher than the surrounding rim. Also, it did not explore the reasons why crater basins are always below surface level and not at various heights right up to the crater rim. Although Naysmyth and Carpenter's arguments were faulty, theirs was a first stab at a systematic understanding of selenophysics.

Richard Proctor (1837–1888), in a book called *The Moon* published in 1873, concluded that craters were caused by material from elsewhere hitting the Moon. What might cause these impacts? The obvious candidates were **meteoroids,** rocky debris wandering through the solar system, which might be captured by the Moon and crash there. The problem was, no one had seen this happen on the Earth, much less the Moon, and the Earth was a larger target. The protective nature of the Earth's atmosphere, which causes incoming meteors to heat because of friction and to be destroyed, was not adequately considered.

Meanwhile, there were additional competing theories for the origin of lunar craters. One said that cratering was caused by repeated straining of the lunar surface as a result of the strong tidal influence of the Earth. Another theory suggested that craters were the remains of burst bubbles that arose out of the early, liquid lunar surface.

The impact theory, though, continued to gain ground. In 1892 an American geologist, Grove Gilbert (1843–1918), explained central peaks in craters as a reaction against impacting meteoroids. They were the solidified recoil to the blow that the surface received. Gilbert went on to say that rilles were cracks formed by impacts and that rays are jets of material thrown out by meteroroids at impact. Today the former is suspect, but we look at the latter as a good explanation of what happens to form lunar rays.

Rays are good evidence supporting impact cratering. So are small craters that often cluster around large craters. These **secondary craters** can be explained as having been formed by the impact of chunks of material hurled up by the impact of a meteoroid. Even tertiary craters are possible.

Impact origin explained the morphology of lunar craters so much better than did volcanism that this theory would have dominated sooner except for the nagging question: If there are so many impact craters on the Moon, why aren't there any on Earth? In 1895, Gilbert stated that there had been many impact craters on Earth during its long history but that wind and water conspired to erase evidence of them.

What if an impact had occurred recently? It would not have had time to erode away yet. The odds for a large impact in recent times are not good, but it has indeed happened. In 1891 a crater was discovered in northern Arizona. It was at first thought to be volcanic. A mining engineer and geologist from Philadelphia, Daniel Barringer (1860–1929), recognized in it distinct nonvolcanic features that pointed to an impact origin. Sure enough, iron meteortes were found scattered around the crater.

Barringer saw that if a large meteoroid had struck in the Arizona desert within recent geologic time, it would stand a good chance of being well preserved today. He began drilling for what he supposed would be a large remnant body buried underneath the crater. He didn't find anything at first. Later, he noticed that rock strata along one side of the crater appeared to be lifted. Barringer concluded that a meteoroid striking the ground obliquely had deposited itself to one side of the main crater. Here, Barringer's drill did hit metal-like material and stopped. Modern probing techniques have identified the presence of a higher density region under the rim.

Today, the Barringer Crater is considered to be the best preserved impact crater on Earth. It is not the only one, however. Others have been recognized in Texas, Australia, and Germany.

The Earth does suffer meteoroid impacts. Fortunately, such events are rare. Still, given enough time and the absence of erosion (as on the Moon), they could accumulate. Thus, one of the major objections to the impact origin theory for lunar craters has been overcome here on Earth.

Why, though, did Barringer not find a solid body implanted in the ground? When a large body intercepts the Moon (or the Earth), the amount of energy that is released may be a million times greater than the yield of a ton of conventional chemical explosives. The behavior of materials under these circumstances is different from what one might expect. In 1921, Alfred Wegener (1880–1930), the father of the theory of continental drift, examined this phenomenon in *Die Entstehung Der Mondkrater (The Origin of Lunar Craters)*. Wegener showed that when a massive object falls to the Moon with a velocity equal to or greater than escape velocity, it strikes with a force that makes the internal molecular forces holding the object together insignificant. In order to model such an impact, Wegener purposely chose materials that have no tensile strength. Specifically, he dropped cement dust into a layer of the same powder. The result of this simple experiment was the production of craters. Wegener's craters were of a size (measured as the ratio of the depth of a crater to its diameter) between that of the Barringer Crater and typical lunar craters. Their rims and floors looked very much like those of craters found on the Moon.

Next, Wegener used plaster of paris as an impacting body. He found that the resulting craters formed with a smooth layer of plaster of paris on their bottoms. Additional plaster of paris was piled up at the crater edges. Some even spilled out in a pattern suggesting rays. In no case was there a single, central remnant of the impact body. Rather, it was spread about the crater as debris, particularly at the rim. This is exactly what Barringer found in his crater! In the Barringer Crater iron-bearing meteoric rock is spread around with a slightly higher concentration on the side aligned with the original body's oblique trajectory.

Wegener explained that the disintegration of the incoming body is what limits the depth of impact craters and causes them to be distinctively wide and shallow. As a body hits, material spreads outward radially from the point of impact. It piles up when this wave disturbance dissipates and forms a rim. In other experiments, Wegener was even able to demonstrate how central peaks occur.

Since the mid-twentieth century, above-ground nuclear explosions have mimicked the meteoritic impact mechanism. Scientists studying this process can see the differences in the depressions formed by increasingly large increments of energy by observing the craters formed in this way. The analysis of crater formation is, perhaps, the sole beneficial use of such explosions.

Impacts account for most of the lunar craters, but there is still a minority that undeniably appear to be volcanic. These include some small, rimless craters that may be volcanic vents and halo craters with dark circles around them (thought to be volcanic ash). In addition, there are too many craters situated atop domed structures to be accounted for by coincidental impacts alone. Finally, all craters associated with tectonic features are suspect.

Is the Moon volcanically dead today? Not necessarily. In recent times, there have been some notable claims of observation of activity on the Moon. In 1958 Nokolai Kozyrev (1908–1983) of the Pulkovo Astronomical Observatory in the U.S.S.R. saw a brightening around a crater. He was able to obtain a spectrum of it, which he claimed showed the presence of carbon-containing gases. Barbara Middlehurst has catalogued this case as well as other reports of **Transient Lunar Phenomena** (TLPs) on the Moon. These include some in or near the large crater Aristarchus. Middlehurst has found that these events tend to occur more frequently near lunar apogee and perigee, times of greater tidal strain.

Brightening or the presence of gas does not necessarily imply the presence of volcanoes. Trapped gas escaping from lunar rock can produce dust. If dust were stirred up by such outgassing as it was illuminated by sunlight, a brightening might be observed. Alternately, there may be electrical discharges associated with outgassing.

The theory of the development of the lunar maria

follows lines similar to those of the theory of the origin of craters. This may be surprising because the two types of features seem very different. This difference, however, is largely a matter of scale. The maria basins are round depressions just like craters. They are the result of a few very large impacts.

One theory had bodies impacting with enough energy to break the Moon's crust. Liquid lava beneath the crust would then ooze forth. It would spread out over the impact area, obscuring existing craters, rays, and other features. The result would be a large, flat region—in other words, a mare. Eduard Suess (1831–1914) modified this theory by suggesting that the impacts did not crack the crust but instead melted it. The heat of impact, he theorized, would conduct radially outward until it was dissipated, again forming a large, flat region.

Today elements of both theories have been combined to form the following scenario: A large impact excavates a mare basin. There is only local melting, however. For a period, the basis remains a lunar pockmark. Subsequently, lava wells up through the depression. This happens in successive stages. Each stage leaves a sheet of new mare material down on the basis floor until a smooth solidified "lava sea" is formed. The "wound" is healed. Eventually much of the lower regions on the Moon are inundated.

Evidence of widespread flooding over older terrain includes **ghost craters.** Ghost craters are the rims of old, partially submerged craters that stick up above maria. Some craters were totally flooded or smoothed over by lava, and only their central peaks remain. Mare formation, lava flooding, and most crater impacts occurred early in the history of the Moon. In Chapter 8 we discuss a scenario for the evolution of the Moon's face. Except for possible TLPs, this evolution is no longer evident to us. On humankind's short time scale the Moon is changeless: The lunar surface that we saw on television in the 1960s is indistinguishable from that viewed through primitive telescopes in the 1660s.

6
Getting There

THE ROCKET

On and off we keep encountering the name of Johannes Kepler and his contributions to lunar science in the realms of celestial mechanics, optics, and selenography. At the end of his life, he published one more major work that dealt with the Moon. Kepler's *Somnium* (1634) is a story about a fictitious trip to the Moon. As early as A.D. 165, the Greek satirist Lucian of Samosata had written about a voyage to the Moon (his protagonist's ship is caught up in a water spout and literally sails to the Moon), but Kepler's tale was not just fantasy. He attempted to describe conditions on the Moon in terms of the scientific understanding of the day. Therefore, he deserves credit as the author of the first science fiction novel about the Moon.

In *Somnium,* Kepler's hero is transported to the Moon in a dream. A story published in 1638 by Bishop Francis Godwin told of a man carried to our satellite by a flock of wild swans! In 1656, Cyrano de Bergerac wrote a story (published posthumously) about a lunar trip that dealt with the method of getting there in a more concrete fashion. An intrepid explorer of de Bergerac's story rides to the Moon by sitting in a chair with **rockets** strapped onto it. This particular method would be, in practice, somewhat dangerous, but it is not far from the image we have today of an astronaut sitting in his **spacecraft** couch atop a giant rocket booster.

Why the rocket? De Bergerac must have seen rockets; they had been around a long while since first making their appearance in China in the 1200s. There they were used for entertainment and ritual and also began their long history as devices of war (rocket-powered arrows, for example). In the nineteenth century, rockets were perfected in this latter role by Europeans, preferably as a naval weapon. (In the National Anthem of the United States we sing, ". . . The rockets' red glare . . .")

Rockets, though, never superseded the cannon and were soon deemed unreliable as gunnery improved. By the time Jules Verne wrote his now-classic 1865 novel *De la Terre à la Lune (From the Earth to the Moon,* the logical way to imagine flying to the Moon was in the projectile of a nine-foot-caliber cannon. Verne prophetically placed this device on the Florida peninsula and made several other insightful guesses as to how people would one day venture to the Moon in reality. The concept of the Moon cannon prevailed well into the twentieth century. In the first true science fiction motion picture, *Things to Come* (1936), two young people are "shot" to the Moon.

Journeys to the Moon were the subject of fiction only, of course. At least this was so in most people's minds. Yet there were a few farthinkers who actively pursued that dream. The British Interplanetary Society and the German Society for Space Travel did pioneering work in making space travel a reality. These enthusiasts realized that firing people into the sky in giant mortar shells was impractical. The shock of the initial acceleration necessary to reach the Earth's escape velocity would be too great to withstand. They placed their hopes in the rocket once again. Their mentors were rocket pioneers like Konstantin Tsiolkovsky (1857–1935), Hermann Oberth, and Robert Goddard (1882–1945). The advantage of the rocket was that it could provide *sustained* thrust over a period of time. Thus, the necessary speed could be built up gradually.

Many people believe that the rocket works because its exhaust "pushes" against the air. This is not true, and if it were, it would certainly make the rocket worthless in the vacuum of space. The rocket works because of Newton's Third Law, simply stated as: "For every action there is an equal and opposite reaction." Let's see how this works.

Imagine that you are in a small boat on a lake and have been totally becalmed some distance from the dock. You have no oars on board and do not feel like "dog paddling" with the boat in tow. Now suppose that you also have a quantity of bricks in the boat (that you don't mind getting rid of). You remember the Law of Conservation of Momentum and proceed as follows: You pick up a brick and toss it as hard as you can in the direction *opposite* that of the dock. In this way, you impart velocity to the brick. Previously, the momentum of your system was equal to zero (you, the bricks, and the boat were all stationary). You and the boat now move toward the dock to make up for the brick moving away from it. Newton's "action" is the flying brick; his "reaction" is the boat moving.

To conserve momentum, the momentum of the boat must be equal and opposite to that of the brick so that the sum of the two momenta equals zero. This means that because the boat is much heavier than the brick, the speed that it gains to "cover" the brick's momentum is much less than the brick's. You may hurl the brick as hard as you can, but the boat will move only slowly. In a short period of time, the resistance of the water and air will bring your boat to a standstill again, but at least you are closer to the dock than you were before.

But there are still bricks left! You pitch another one away from the boat, and your vessel moves again. As you keep throwing bricks behind you, you eventually propel your boat to the dock (assuming that you don't run out of bricks). You arrive there courtesy of the Law of Conservation of Momentum, successfully completing your voyage, unless, of course, your object was to transport a pile of bricks to the dock!

Imagine that instead of a boat and a pile of bricks we have a rocket and its exhaust. Inside a rocket engine, gas is heated by combustion. This gas has a tremendous amount of kinetic energy and escapes at a high velocity out of a nozzle at the rear of the rocket. The mass of each gas molecule is miniscule compared to that of the rocket, but owing to their high speeds, they carry with them a significant amount of momentum. Also, at any instant there are trillions of molecules escaping from the rocket. It is this continuous outpouring of momentum that propels a rocket forward. An action (gas rushing out of a rocket engine) has a reaction (the rocket moving forward).

A rocket need consist only of fuel, an engine to burn the fuel and produce a hot expanding gas, and, if it is to operate in the atmosphere, an aerodynamically designed outside shape to allow it to slide smoothly through the air. (This is why most rockets that we see are smooth and tapered at the top.)

The problem with using rockets for transportation was that they used a solid fuel (like bottle rockets use today). Once the fuel was lit, it would continue to burn, and the rocket would continue to fire uncontrollably until it was out of fuel. In the 1920s Robert Goddard invented a liquid fuel rocket. Not only was liquid fuel more efficient, its flow into the engine could be controlled, thereby providing a throttle on its performance and allowing it to be turned on and off at will. The development of the liquid fuel rocket was a major step toward making it into a vehicle for space travel.

In England the enthusiasts who experimented with rockets were branded as eccentrics. In Germany, however, during the 1930s many of them were put to work developing a rocket for quite a different purpose from the exploration of the solar system. The resulting V–2 rocket was capable of delivering not an arrow across a battlefield but a bomb across countries. (Figure 6–1). It was a dramatic (if ineffective) weapon that Hitler let loose against Great Britain during World War II.

At the close of the war, a race began between the

(U. S. Army photo)

6–1 A V–2 Rocket being readied for launch at White Sands, New Mexico.

United States and the Soviet Union—a race to get hold of German rocket technology. The Russians captured the German rocket base at Peenemünde, but most of the German rocket scientists fled to the West. They were quickly ensconced at Ft. Bliss, Texas, and later at Huntsville, Alabama, to develop an American rocket program. The goal was to marry the rocket to another product of the war, the atomic bomb. There seemed to be plenty of time. The Soviets were considerably behind in the development of nuclear weapons. Ironically, because the crude Soviet bomb was so heavy, it necessitated the building of a large launch vehicle, which the USSR did with an alacrity that shocked the western world.

With the development of a large Intercontinental Ballistic Missile (ICBM), only a relatively simple modification was required to carry a nondestructive payload on a one-way trip. The booster would be used to reach **orbital velocity** and place a second "moon" circling about the Earth. This was accomplished with the successful launch of Sputnik I, the world's first artificial satellite, on October 4, 1957. It was followed by Sputnik II on November 3 (carrying a live dog as a passenger).

The Soviet feat caused great excitement in the United States and near-panic among government officials and scientists who had been methodically developing this country's own artificial satellite. It didn't help when the first American attempt, Vanguard, blew up on its launch pad on December 6, 1957.

If the Russians could put a satellite into orbit, would a bomb be next? The image of a Soviet warhead passing 100 miles above American heads several times each day was conjured up. Clearly, there was a space race taking place and the United States was losing.

The Army, using a modified ballistic missile, quickly placed an American satellite in orbit on January 31, 1958. Similar in concept to Vanguard, Explorer I did little more than beep out a radio signal as Sputnik had and weighed only 8 kilograms (compared to 84 kilograms and 510 kilograms for Sputniks I and II respectively).

On August 19, 1960, the USSR sent two dogs to an altitude of 340 kilometers and recovered them safely. Was Russia planning to send people into space? The goal of being first to put a person into space became the next test of who was where in the space race.

Once again, it was to be the Soviets who accomplished the job first. On April 12, 1961, Yuri Gagarin became the first man in space. He made a single orbit of the Earth, completing his mission in 1 hour, 48 minutes.

The United States launched Alan Shepard into space three weeks later. Shepard's flight on May 5 was the first in a series of Project Mercury missions designed to develop manned space flight in a step-by-step manner. The first flight and another on July 21 made by Virgil Grissom were **suborbital** flights. The astronauts went up and down without achieving orbit. An American orbital flight was not flown until February 20, 1962, when John Glenn made three revolutions in 4 hours, 55 minutes. (Gherman Titov of the Soviet Union had already spent more than a day in space aboard Vostok II.)

The United States seemed to be perpetually behind in space accomplishments. What would the Russians do next? It was during this nadir of confidence in the United States' space program that President John Kennedy told the world: "I believe that this nation should commit itself to achieve the goal before this decade is out of landing a man on the Moon and returning him safely to Earth." The Apollo Program was born.

The quest for the Moon was undertaken by the National Aeronautics and Space Administration (NASA), which had directed Project Mercury. The amount of work that the fledgling agency had to do was enormous. Not only did it have to learn how people could work and function in space; it had to build a launch vehicle big enough to get to the Moon.

There were several ways of putting together such a vehicle. First, a huge rocket could be built, one big enough to take a spacecraft all the way to the Moon and back by itself. An alternative called for several rockets carrying aloft the parts for a moonship. These components would be assembled in Earth orbit. A compromise between these two plans was adopted. A large rocket would put two separate spacecraft into Earth orbit. It would later reignite to send them to the Moon. Each of the two spacecraft would be designed for a specialized task. One, a command module (CM) and service module (SM), would house the astronaut crew on the way to the Moon and then return them to the Earth. The other, a lunar module (LM), would be designed specifically to land on the Moon and would be disposed of after its job was through.

NASA called upon Wernher von Braun (1912–1977), a veteran of the German V–2 program, to design the new booster on top of which the CM/SM and LM would leave the Earth. Von Braun would build the largest rocket ever to travel into space. The rocket would be 110 meters tall (roughly the height of a 30-story skyscraper) and nearly 100 times as powerful as that used to launch Explorer I. It was to be called the Saturn V–the Moon rocket.

PROJECT GEMINI

Project Mercury continued. Scott Carpenter flew a mission identical to John Glenn's on May 24, 1962. Walter Schirra made six revolutions of the Earth on October 3, 1962. Gordon Cooper spent 34 hours, 20 minutes in space on May 15 and 16, 1963. He was the first person to sleep in space.

Meanwhile, the Russians made another space "first" by flying the Vostok III and IV missions simultaneously in August 1962. Vostoks V and VI flew in formation in June 1963. The pilot of Vostok VI was the first woman in space, Valentina Tereshkova.

The Soviets had a monopoly on space from mid-1963 through 1964. During this time, they introduced a larger spacecraft called Voskhod, which carried two- and three-person crews. During the flight of Voskhod II (March 1965), Alexei Leonov became the first person to take a spacewalk while conducting 10 minutes of Extravehicular Activity (EVA).

Where were the Americans? During this intermission, NASA was preparing for a new series of space missions designed specifically to pave the way for Apollo. Project Gemini would fly two-person crews in a larger spacecraft. This new vehicle would be maneuverable and outfitted for flights of longer duration.

Mercury had established that people could function in space, and Gemini would attempt to prove that they could perform useful work there—tasks necessary for the successful completion of the projected Moon flights. Gemini would test procedures for **rendezvousing** and **docking** spacecraft, both techniques vital to Apollo. It would practice changing both the orbit and the altitude of a craft in space as well as controlled re-entry into the atmosphere. It would measure a person's ability to stay in space for two weeks, as would be required for an extended Moon mission. It would establish a capability for an astronaut to work outside of a spacecraft. Last, it would allow development of procedures for flight and ground crews in manned spaceflight. It would do all of this in 10 missions flown in less than 20 months.

Gemini III left Cape Kennedy, Florida, on March 23, 1965, for space (Figure 6–2). This spacecraft was dubbed

(NASA photo)

6–2 The launch of Gemini III.

the "Unsinkable Molly Brown" by its commander, Virgil Grissom, who had lost his Mercury spacecraft to the bottom of the Atlantic Ocean after landing.

The Gemini was lifted into orbit by a modified ICBM booster named the Titan. The Titan was a multistage rocket—that is, two rockets, one on top of the other. After the first stage was exhausted, its dead weight dropped away. The now lighter rocket was propelled into orbit by the ignited second stage. This orbit was successfully modified during the mission by the firing of an on-board engine and side **thrusters.**

After 3 orbits and 4 hours, 53 minutes of flight, Grissom and spacecraft pilot John Young had flight-tested the new vehicle, pronounced it spaceworthy, and returned to the Earth.

The following June, James McDivitt and Edward White made 62 revolutions around the Earth aboard Gemini IV. White spent 36 minutes during one of these conducting the first American extravehicular activity. The EVA was necessary for Apollo because, in the event that the two Apollo spacecraft could not be joined in space, a spacewalk would be necessary in order to transfer crew members from the returning lunar craft to the Earth-bound one. Also, for astronauts to be of much use on the lunar surface, they would have to have the capability to leave their LM and work outside. The protective suits they would wear on the Moon would be very much like those worn on a Gemini EVA.

The EVA is possible because of this remarkable device, a **spacesuit** that is a self-contained environment for an astronaut. It must mimic all the functions of a spacecraft but on a smaller scale. The suit must regulate air coming from a hose attached to the spacecraft and get rid of heat produced by the astronaut. It carries a backup oxygen supply in case the hose supply fails. It has to withstand humidity, vibration, explosive decompression, acceleration, oxygen exposure, extreme temperatures, and hazards of the space environment. It needs to fit and allow the astronaut to move.

Perhaps most important, it must be something that an astronaut can stand to wear for hours at a time.

The prototype Gemini spacesuit was a direct ancestor of the remarkable Apollo moon suit. Outside Gemini IV, White, encased in his suit, floated about at the end of a long tether and umbilical hose (Figure 6–3). Because both he and the Gemini spacecraft were in free fall around the Earth, White was effectively weightless. To move relative to the spacecraft he pushed against it or operated a small handheld thruster. Even with the thruster, White found that controlling his position and going where he wanted to go were not easy when common-sense Earth concepts such as "up" and "down" had lost their traditional meanings.

Gordon Cooper and Charles Conrad flew the first extended Gemini mission aboard Gemini V: 190 hours, 56 minutes.

(NASA photo)

6–3 White's Gemini IV EVA over New Mexico.

Gemini VI was designed to try rendezvous and docking procedures. During Apollo, it would be necessary for the LM and CM/SM to find, rendezvous, and dock with each other. This had not been practiced in space where conventional pilots' wisdom ends and Kepler's Laws take over (Figure 6–4). For instance, suppose an astronaut wanted to catch up with another craft ahead of and below him. Instinctively, the pilot would simply fire his rear engines. Such an act, though, would *raise* his orbit and cause him to fall even farther behind. To lower his orbit and speed up, the astronaut had to fire **retrorockets** forward!

Gemini VI was to practice the consequences of orbital mechanics on spaceflight with an unmanned Agena target vehicle, which was launched just before Gemini VI's scheduled liftoff. The rocket carrying the Agena malfunctioned, and the target vehicle was lost. The Gemini VI crew, Walter Schirra and Thomas Stafford, were all dressed up with no place to go. The flight was scrubbed. (Figure 6–5). A

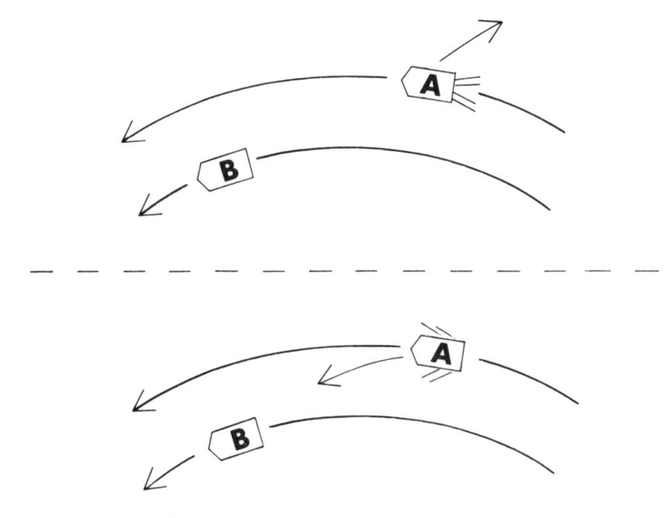

6–4 Orbital Rendezvous. Spacecraft A is attempting to rendezvous with spacecraft B. *(Above)* A fires its engine, thereby increasing its orbital altitude, slowing, and falling farther behind B. However, by firing in the opposite direction *(below)*, A loses altitude, speeds up, and overtakes B.
(Andrea K. Dobson-Hockey)

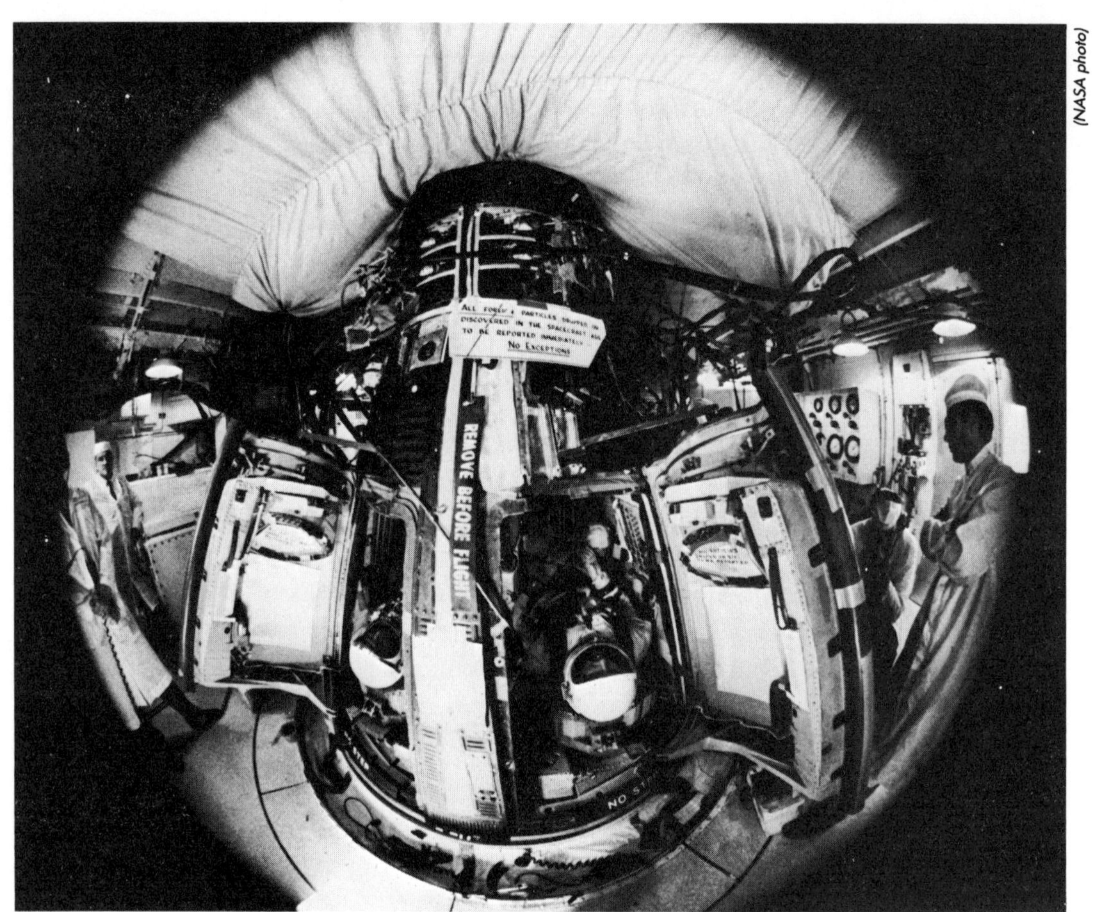

(NASA photo)

6–5 Gemini VI before launch. Schirra is in the left seat; Stafford is in the right.

quick change of plan was called for. Gemini VI was delayed until *after* the launch of Gemini VII. The launch times were synchronized so that Schirra and Stafford could use Gemini VII as their target vehicle for rendezvous practice. (Gemini VI underwent a second launch scrub before lifting off successfully on the third try.)

Gemini VI maneuvered around and behind Gemini VII, allowing each crew to inspect the other's craft. At times they approached to within a meter of each other. By the time Gemini VI and VII had parted company and Schirra and Stafford had returned to Earth, the American space program had achieved near parity with that of the Soviets.

The Gemini VII mission continued. Launched 11 days before Gemini VI, it did not return to Earth for 2 more days. Its crew, Frank Borman and James Lovell, spent a record two weeks in space. Their mission goals were primarily medical—to test astronauts' ability to spend lengthy periods of time in space.

The mechanics of maintaining people comfortably in the hostile environment of space for days or even weeks are difficult to begin with. A spacecraft needs to be a miniature reproduction of the Earth. It provides oxygen for the crew to breathe, processes out the carbon dioxide they exhale, keeps them from getting too hot or too cold, maintains air pressure, and manages waste. It also stores food, produces electricity with **fuel cells,** and provides a communications link with the Earth. It is vital that a spacecraft perform these functions reliably when it becomes the astronauts' only haven on their journey to the Moon. Despite minor thruster and electrical problems, the Gemini VII crew completed their mission and demonstrated the capability for an extended flight (330 hours, 35 minutes) (Figure 6–6).

Gemini VIII successfully rendezvoused and docked with an Agena vehicle. However, the exercise was terminated when a stuck thruster set the joined spacecraft and Agena tumbling end over end. This precipitated an emergency undocking and an early landing in the Pacific. Crew members Neil Armstrong and David Scott did not have the opportunity to use the Agena propulsion system to boost them to a higher orbit, nor did Scott have the chance to make the intended spacewalk.

Gemini IX experienced troubles, too. Its launch was postponed twice: once because of the failure of another Agena target vehicle and once because of computer problems. After all that, a protective housing covering the substitute target vehicle docking mechanism failed to open all the way after reaching orbit. Its faulty shroud gave it the nickname, the "angry alligator."

Once Gemini IX was finally launched, Commander

(NASA photo)

6–6 Lovell *(left)* and Borman on the USS *Wasp* after their fourteen-day mission aboard Gemini VII.

Thomas Stafford and Pilot Eugene Cernan were able to practice rendezvous maneuvers with the angry alligator that simulated an abort of a Moon landing mission during descent to the surface.

Cernan also went for a spacewalk. He tested handholds and foot restraints installed around the exterior of Gemini IX that were designed to help him keep his position and move around the spacecraft. These proved to be inadequate, and Cernan expended a great deal of extra energy completing his tasks. His helmet visor finally fogged, and the EVA was ended prematurely after 2 hours, 7 minutes.

Gemini X was more successful. Crew members John Young and Michael Collins successfully docked with its target vehicle and used the Agena's engine to boost them to an orbit with a record 412 kilometer-high apogee. Gemini X also rendezvoused with the four-month old Gemini VIII Agena.

Still, though, spacewalking was not as easy as it

seemed. The goal of Collins' EVA was to retrieve experiment packages from the Gemini and Agena. This was accomplished, but too much fuel was expended keeping the two vehicles in formation, and the EVA was terminated after 39 minutes. Returning to the Gemini, Collins became entangled in his umbilical cord. A 70 millimeter camera also managed to float away. To add insult to injury, during a subsequent hatch opening one of the experiment packages was thrown out with the trash.

Gemini XI far surpassed Gemini X's altitude record. With the help of an Agena, it reached a height of 1,368 kilometers. At this distance from Earth, astronauts Charles Conrad and Richard Gordon were continually monitored for effects from the **Van Allen radiation belts** that surround the Earth and pose a potential hazard for Moon-bound Apollonauts. No excessive radiation was detected.

More footrests and restraints were in place for Gordon's spacewalk, but he found himself overexerting to keep in place and accomplish his tasks. The EVA was concluded after only 33 minutes.

The last of the Gemini series, Gemini XII, with James Lovell and Edwin "Buzz" Aldrin aboard, was also the most successful. Its primary goal, a satisfactory EVA, was achieved. By installing new and better holds (including molded foot restraints) and programming sufficient rest breaks into Aldrin's EVA schedule, he was able to complete his objectives. He spent 2 hours, 8 minutes working outside his vehicle.

The whirlwind Gemini program did well in accomplishing its goals. In the untested field of rendezvous and docking where all technology and techniques had to be developed from scratch, Gemini completed 10 rendezvous, accomplished in 7 different modes, and nine different dockings.

EVAs were performed on five Geminis. It was shown that, given an understanding of how people work in a weightless state, useful work can be performed outside the spacecraft.

Re-entry and recovery procedures were honed and the range of orbits and altitudes that a manned spacecraft might operate in was broadened. Almost incidentally, Gemini astronauts conducted 19 scientific experiments and took more than 2,400 photographs.

Perhaps Gemini's biggest contribution was in improving knowledge of the medical aspects of space flight. The crews withstood the high accelerations of liftoff and reentry, which caused them to feel a force of up to eight times their own body weights. It was possible to maintain the Gemini at a comfortable temperature and air pressure. The rapid fluctuations between day and night in orbit (once every hour and a half) did not unduly disrupt the astronauts' sleep patterns. (Shades on the windows helped!) The danger of **micrometeoroids,** small projectiles in space, penetrating the spacecraft was found to be minute. Radiation shielding was adequate.

The astronauts suffered no harm from being cooped up in their spacecraft for up to two weeks. Effects of isolation and confinement that had been predicted—hallucinations, space sickness, lethargy—did not materialize. In fact, the major fatigue problem seemed to be caused by the astronauts' being tempted to stay up too late watching the spectacular view out of their windows!

Research and development of the capability to go to the Moon was the cornerstone of Gemini. This often-overlooked program was vital to Apollo. Gemini was all the more remarkable in that its successes and failures were all completed out in the open under a media spotlight for all the world to see. This unique policy remained a feature of all U.S. manned spaceflights until the launch of defense-related mission STS–51–C in January 1985.

RANGERS, SURVEYORS, AND ORBITERS

Not only did NASA have to create the technology necessary for taking people to the Moon, it had to learn something about the place it was sending them. Lunar science had waned in the first part of the twentieth century. The Moon was thought to be fully understood. Astronomers pursued more exotic celestial objects with the large telescopes that had become available. At the inception of Apollo, though, the possibility of looking at another world up close and actually picking up and bringing back parts of it refocused attention on the Moon.

Once again the Russians took the lead. On January 2, 1959, the USSR launched a "cosmic rocket." Atop the booster was a Luna spacecraft—the first to fly by another celestial body. Luna I's path actually took it past the Moon and into an orbit about the Sun. On September 4, Luna II crashed into the Moon, becoming the first man-made object to rest there.

On October 7, Luna III sent back pictures of the lunar Farside. A surprise awaited scientists viewing the images made by this and later spacecraft. The lunar Farside had been expected to be much like the lunar Nearside. This expectation was wrong! The Farside turned out to be almost devoid of maria. The hemisphere resembled the highly cratered highlands. Those impact basins that existed remained exposed and unfilled for the most part. This was the first of many spacecraft discoveries that were to establish how much more complex the Moon is than had previously been thought.

The United States now began to design a **space probe** to explore the Moon. The probe, to be called Ranger, would intentionally crash-land on the Moon. On board would be a camera and an instrument package designed to withstand the hard landing and transmit information about the lunar surface back to the Earth. Ranger was under development when the call to conduct a manned lunar landing was made. It became necessary to proceed with

Ranger with all due haste. The instrument package was slowing spacecraft development down; it was abandoned. Only the camera was kept as a necessary tool for finding out what hazardous conditions might await a manned vehicle on the lunar surface.

The first five Ranger test flights malfunctioned. Then on January 30, 1964, Ranger VI was successfully launched on course for the Moon. It crash-landed there three days later. Unfortunately, the spacecraft cameras failed to turn on. Ranger VI crashed into Mare Tranquilitatis without taking a single picture.

Ranger VII experienced a different fate. It was launched into Earth orbit on July 28. Later its booster reignited to direct it on a trajectory intercepting the Moon. The spacecraft separated from the booster and unfolded its **solar panels** that were designed to supply electricity from sunlight. An hour later sensors locked onto the Sun and the Earth to orient the spacecraft. A small on-board thruster was fired to make minor adjustments in Ranger VII's course.

At a distance of 1,900 kilometers from the Moon (when Ranger was traveling at nearly 7,100 kilometers per hour) the six TV cameras were turned on. The first pictures were taken at a distance of 1,800 kilometers. The results were impressive. In 13 minutes, 40 seconds, Ranger VII sent 4,308 images to the Earth (Figure 6–7). Ranger's last picture (interrupted in mid-frame by the destruction of the camera) was taken a fraction of a second before the space probe hit, at a distance of only 529 meters. Objects as small as one meter in diameter can be made out in it. Suddenly our resolution of the Moon had jumped by a factor of nearly 1,000.

In February 1965, Ranger VIII sent 7,137 pictures back to the Earth, transmitting ten minutes longer than Ranger VII. A month later Ranger IX splattered about the crater Alphonsus (a possible site of volcanic activity described in Chapter V), producing still more images.

To decide if the lunar surface was safe enough to set foot on, it would be necessary to place an automated space-

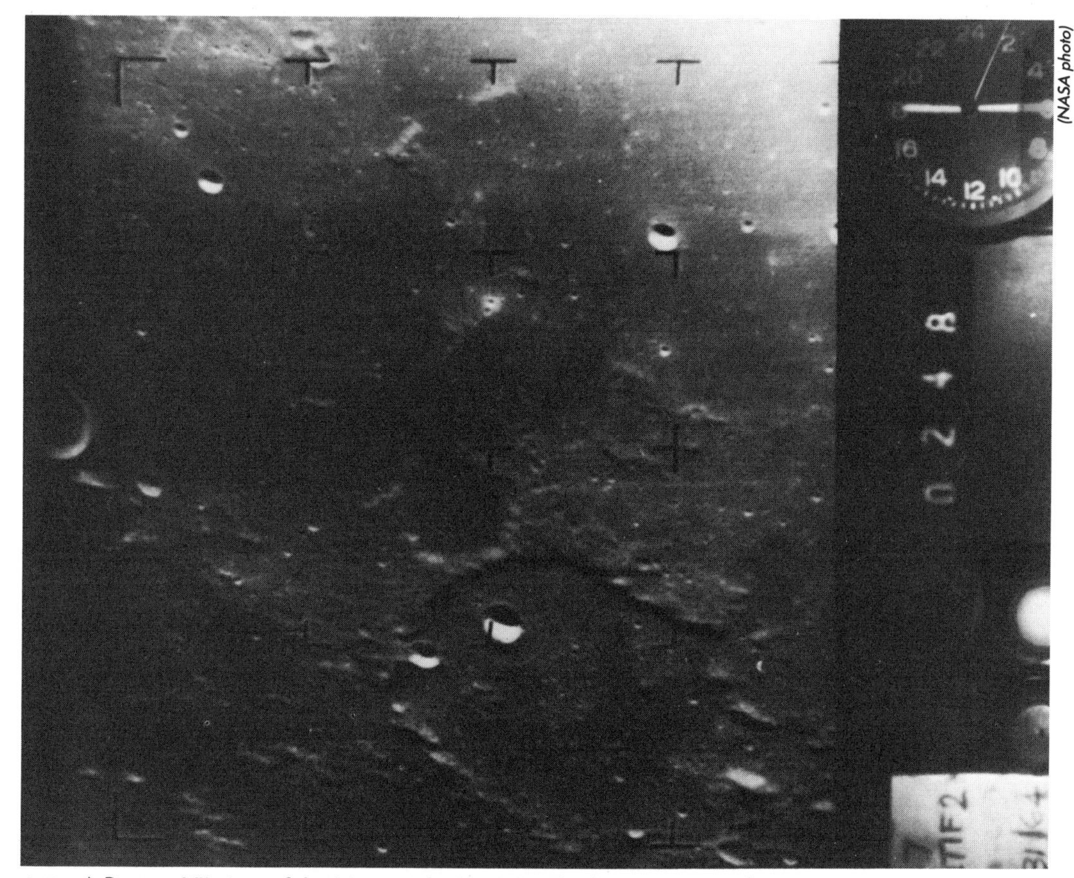
(NASA photo)

6–7 A Ranger VII view of the Moon, July 31, 1964. The large crater in the upper right is Guericke. The smallest craters are less than 300 meters across.

craft there first. Such a "soft" landing was next on NASA's agenda. The Soviets had tried this eight times before successfully accomplishing it with Luna IX in February, 1966. Its landing site was Oceanus Procellarum.

The name of NASA's spacecraft was Surveyor. Surveyor was designed to slow itself as it approached the Moon by firing retrorockets and land safely on three legs. This would allow Surveyor to serve as a station from which to deploy an assortment of instruments. Foremost among these was a camera designed to transmit panoramic views back to the Earth.

Surveyor I was ready four months after Luna IX and was a test flight for the new program. It was a successful mission: Surveyor I hit the Moon at a speed of only 13 kilometers per hour. It also landed in the Oceanus Procellarum.

Surveyor I carried no instruments except its camera. This device performed excellently, however. Pictures began to arrive on the Earth 35 minutes after the landing and were of a higher quality than those of Luna IX. The horizon view from Surveyor I showed the rim of the crater Flamsteed, 83 kilometers in diameter. Up close the camera showed images of rocks as small as kernels of rice. These pictures' resolution was 1,000 times greater than the best Ranger photographs.

All told, 10,386 pictures were transmitted. The spacecraft survived the two-week lunar night and eventually had to be "killed" by turning its solar panels away from the Sun so that it would not interfere with telemetry coming from Surveyor II. Unfortunately, a midcourse correction for Surveyor II failed, and the spacecraft began to tumble. Uncontrollable, it crashed into the Moon at full speed.

Surveyor III narrowly avoided a fatal landing in April 1967. Its retrorockets failed to shut down on time, and the spacecraft bounced three times before coming to rest on the bottom of a crater. The landing site was near the Riphaeus Mountains, again in Oceanus Procellarum.

Besides a camera, Surveyor III carried a robot arm whose mechanical claw dug four trenches in the lunar soil (Figure 6–8). It also deposited a sample of moon dirt onto the Surveyor footpad. Here it was photographed against a color comparison card.

After an uneventful flight, Surveyor IV stopped transmitting just as it touched down.

Launched in September, Surveyor V had a shaky start as well. It slid partially down a crater rim before coming to rest. No damage was done, however. Its landing site was in Mare Tranquilitatis. This Surveyor carried a camera and a device to study the chemical composition of the lunar surface. The device consisted of a small box lowered from the Surveyor to the ground. A radioactive source in the box emitted **alpha particles,** and an instrument detected the

(NASA photo)

6–8 Surveyor III holds a small rock in its robot claw. The rock has been uncovered from beneath the lunar surface.

manner in which the Moon soil responded to them with radiation of its own. In this way the soil provided information on the microscopic particles that make it up.

The flight of Surveyor VI was similar to that of Surveyor V. Surveyor VI landed in Sinus Medii in November. (Figure 6–9). The Surveyor landing sites were chosen by NASA and the United States Geological Survey to provide information about potential Apollo landing sites. With the successful completion of the Surveyor VI mission, the survey was complete. The last Surveyor was a strictly scientific mission. It was decided to risk landing it away from the smooth mare where the other Surveyors had touched down. Its target was the rough lunar highlands where a different type of lunar material could be examined.

(NASA photo)

6–9 A mosaic of the lunar landscape made from Surveyor VI photographs. At the horizon is a low ridge. The rocks scattered along the ridge are up to half a meter in diameter.

Surveyor VII landed on the edge of the crater Tycho in January, 1968. The spaceprobe deployed an alpha particle-scattering experiment, dug a 30-centimeter trench with a robot claw, and took spectacular pictures. Surveyor VII actually sat on material thrown out from the Tycho impact event. Around it were secondary craters. Boulders were scattered around the landing site, any one of which would have destroyed the Surveyor had the lander had the misfortune to come down on top of it. The information yielded by the Surveyor landings is included in our discussion of the lunar surface layer in Chapter 8. Figure 6–10 shows the locations of Ranger impact sites and Surveyor landing sites.

The plan of selenographers for choosing an Apollo landing site was twofold: first, an indepth analysis of lunar surface conditions at specific sites; second, a large-scale, high-resolution mapping of the Moon. Surveyor fulfilled the first task. The second was taken up concurrently by another spacecraft series, the Lunar Orbiters.

Surveyor had used the same television camera as had Ranger, but Lunar Orbiter required something differ-

(Photograph courtesy of New Mexico State University)

6–10 Ranger impact sites (R) and Surveyor landing sites (S) on the lunar Nearside.

ent. As its name implies, a Lunar Orbiter circled the Moon. It did so rapidly in a low orbit. Pictures made by television cameras are built up relatively slowly and would have been blurred as the spacecraft whisked over the lunar surface. Instead, a Lunar Orbiter's two cameras were of the photographic type and used 70-millimeter film exposed by a rapid shutter. Then, an on-board photographic laboratory developed the film. (Still, the entire Lunar Orbiter payload weighed less than 70 kilograms.) The resulting pictures were slowly scanned by a **photomultiplier** for conversion to electronic signals and for transmission to the Earth at a convenient time. In this way, photographs taken of the lunar Farside could be stored until line-of-sight contact could be re-established with the Earth. (The Soviet Union flew the Zond series of space probes to the Moon starting in 1965. Zond took photographs on a single swing around the satellite and returned them to the Earth for processing.)

Five Lunar Orbiter spacecraft were flown in a 13-month period beginning in 1966, and all vehicles performed to specifications. (Luna X also orbited the Moon in 1966.) By operating Lunar Orbiter at the same time as the Surveyor program, Lunar Orbiter mapping could be used to improve Surveyor navigation.

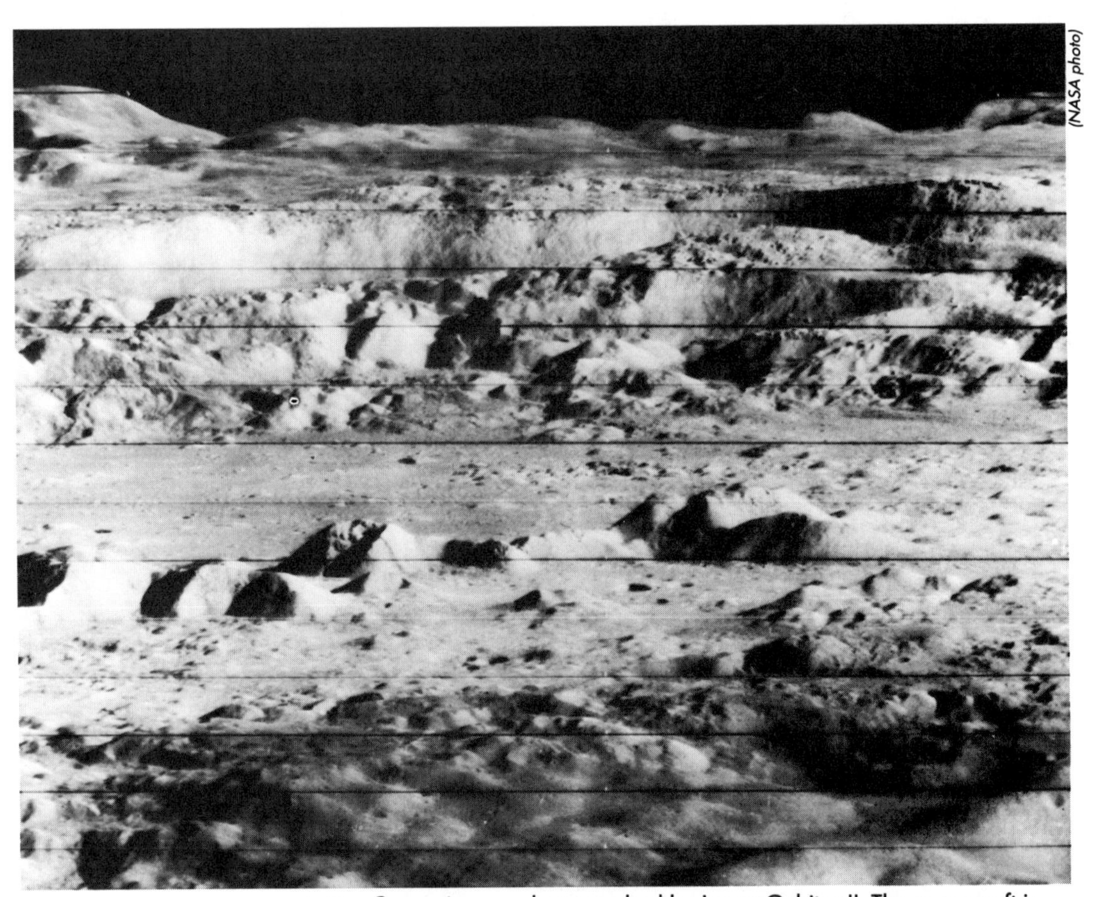

(NASA photo)

6–11 Copernicus as photographed by Lunar Orbiter II. The spacecraft is 250 kilometers from the center of the crater. The mountain in the distance *(upper left)* is part of the Carpathian range and is 1,000 meters high.

Lunar Orbiter I was launched in August 1966. In addition to its valuable Moon photographs, this Orbiter provided the first view of Earth as seen from the Moon.

In November 1966 Lunar Orbiter II took what was at the time called "the picture of the century." Skimming at an altitude of only 47 kilometers over the lunar surface, this spacecraft photographed the spectacular crater, Copernicus (Figure 6–11). Scarps and rays can be seen radiating from the crater in the picture. The crater itself appears to be terraced.

Orbiter II also photographed the Ranger VIII crash site in Mare Tranquilitatis. Lunar Orbiter III photographed the white Surveyor I spacecraft as it sat contrasted against the dark Oceanus Procellarum.

Lunar Orbiters I, II, and III were specifically dedicated to helping to pick Apollo landing sites. Lunar Orbiter IV undertook a broader mapping mission in May 1966. It provided a clear view of Mare Orientale—always partly tucked around the lunar limb when seen from Earth—for the first time. The extent of the concentric mountain ranges surrounding the mare is evident in the photograph. The appearance of **deceleration dunes** (seen at the sites of nuclear blasts and caused by the collapse of a rising cloud of material) testify to the explosive origin of this feature.

Lunar Orbiter V (flown in August) was a scientific mission concentrating on areas of special interest found in the pictures from Lunar Orbiter IV. It took a beautiful photograph of the "Full Earth" never before seen. The high quality of the photographic process used and the low altitude of the spacecraft allowed a resolution of features only about a meter across.

In a little more than a year, Lunar Orbiter mapped the Moon in detail never achieved before. The *Atlas and Gazetteer of the Moon,* produced by NASA, contains pictures of more than 95 percent of the lunar surface at a resolution 10 times that obtainable from the Earth.

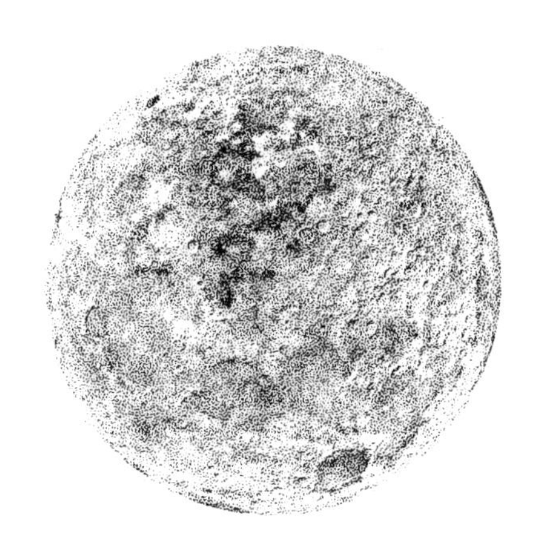

7
THE APOLLO PROGRAM

EARLY APOLLO

Project Apollo began in tragedy. On January 7, 1967, astronauts Virgil Grissom, Edward White, and Roger Chaffee were killed, ironically on the ground, during a test of the Apollo I spacecraft. The optimism that pervaded the manned space program before the accident was wiped out. No American manned space flight would occur for more than 1½ years. Many people doubted that we would get to the Moon on schedule.

When a manned Apollo spacecraft next sat on its launch pad, the date was October 11, 1968. Hope now rested on the redesigned Apollo CM, which the crew of Apollo VII was to test (Figure 7–1). The mission lasted 11 days. Astronauts Walter Schirra, Donn Eisele, and Walter Cunningham put the spacecraft through its paces, and NASA gave the go ahead once again for sending Apollo to the Moon.

The problem was that although the Apollo CM was ready, the LM was not. Worried that the Soviets would soon make their move, NASA decided to send Apollo VIII on a lunar orbiting mission without a lander. At least, the Americans would be the first to go to the Moon if not the first to set foot there.

Frank Borman, James Lovell, and William Anders were the first astronauts to be rocketed into space atop the mighty Saturn V. After they checked out their vehicle in

(NASA photo)

7–1 Apollo VII launch.

Earth orbit, they received the signal to reignite the Saturn third stage—the S4B—for translunar insertion. This engine burn gave Apollo VIII escape velocity and sent it on its historic course for the Moon.

During the part of the mission known as the translunar coast, the astronauts took pictures of the receding Earth. Apollo VIII passed the mythical boundary between the Earth and the Moon—the point in space at which the gravitational force felt by a spacecraft from each body is equal. Now attention was focused on the Moon. On December 24, Apollo VIII looped behind the satellite. The single

(NASA photo)

7–2 The rising Earth as seen from lunar orbit. This popular Apollo VIII photograph later inspired a United States postage stamp.

SM engine was fired to put the spacecraft into lunar orbit (Figure 7–2).

Immediately the crew started to report their impressions of the Moon. Lovell: "essentially gray, no color, looks like plaster of paris or sort of grayish beach sand. . . . it makes you realize just what you have back there on Earth. . . ." Then Borman: "It certainly would not appear to be a very inviting place to live or work. . . ."

In their famous Christmas Eve broadcast via television, Borman, Lovell, and Anders beamed back to the Earth closeup pictures of the Moon. The men's reading from the Book of Genesis would be remembered long after the Apollo program came to an end. Later that night, a critical SM engine burn was made. A malfunction would have left the astronauts permanently stranded. After the successful trans-earth insertion Borman said, "There *is* a Santa Claus."

The Apollo VIII re-entry was the fastest return to that time. The CM (the SM was jettisoned late in the mission) fell to the Earth from the Moon with its blunt end forward to absorb and ablate the tremendous heat generated by atmospheric friction. The spacecraft splashed down in the Pacific Ocean at sunrise on December 27.

Apollo IX was in some sense anticlimactic after Apollo VIII. After all, it was to stay in Earth orbit. However, in other ways it was more important because Apollo IX carried aloft the first lunar module to be test flown by men in space. Liftoff occurred on March 3, 1969. Once in orbit, astronaut David Scott separated the Apollo CM/SM from the Saturn S4B, turned it around, and docked with the LM, which sat directly above the third stage. He then pulled the lander out of its housing, following the same maneuver that moon-bound Apollonauts would use during the translunar coast. The most important part of the mission occurred on March 7. Russell Schweickart and mission commander James McDivitt separated the LM from Scott in the CM and flew it as an independent spacecraft. The astronauts were totally dependent on rendezvousing and docking with the CM again because the LM was designed exclusively for operation in space, and its flimsy construction would not withstand re-entry. During the time that the two spacecraft were separated, they used the nicknames *Gumdrop* (CM) and *Spider* (LM).

Apollo X was to be a dress rehearsal. The time for the first landing was rapidly approaching. Apollo X would simulate a lunar landing mission in every way except the actual touchdown on the Moon. On May 8, 1969, Apollo X followed the route forged by Apollo VIII. During the translunar coast, CM Pilot John Young docked with the lunar module just as David Scott had on Apollo IX.

Once in lunar orbit, the Apollo X crew described the Moon as "fantastic," "incredible," and "unbelievable."

On May 22, Commander Thomas Stafford and LM

Pilot Eugene Cernan crawled into the LM named *Snoopy* and separated from Young in the CM *Charlie Brown.* They descended toward the lunar surface. At their closest point they were only 14,300 meters above the Moon. Tantalizingly near, the men now abandoned their trajectory and left the feat of landing for someone else. Stafford and Cernan jettisoned the LM descent stage and fired the ascent stage for the trip back up, thereby simulating an aborted lunar landing.

The remaining Apollo flights were to be more than astronautical record-setters. They were voyages of discovery. In the next several sections, we follow the events of these flights as stories of penultimate exploration.

APOLLO XI

The time had come for the event itself. The expense, the difficulties, and the years of labor were behind Project Apollo now. The flight was Apollo XI. The crew consisted of Gemini veterans Neil Armstrong, Edwin "Buzz" Aldrin, and Michael Collins. The astronauts believed that historic names were appropriate for a historic mission, names that would reflect the United States itself. The designations they chose for the CM and LM were *Columbia* and *Eagle,* both powerfully symbolic names in this country.

Wednesday, July 16, 1969, dawned bright with only a little haze. More than a million people crowded around the Cape Kennedy area, including 3,000 reporters and 15,000 VIP guests. Many others had camped on the shores and beaches surrounding the moonport to get a good view of the fleeing rocket.

At 9:32 A.M. EDT, Apollo XI rose into the sky to meet the late President Kennedy's challenge. It orbited the Earth for 2½ hours at 190 kilometers altitude and then reignited the S4B to send the CM/SM on its course. (Figure 7–3). *Columbia* separated and docked with *Eagle* in the S4B. That evening, the crew opened the LM to check for damage or malfunction. Apollo XI crossed the midpoint between the

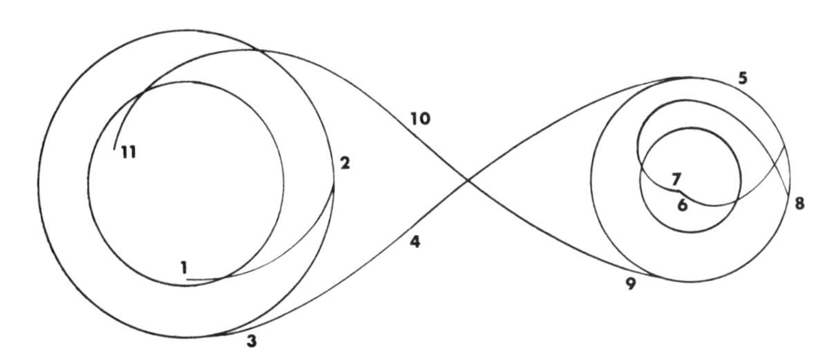

7–3 A summary of an Apollo lunar landing mission: 1) launch 2) Earth parking orbit 3) Trans-Lunar Insertion 4) Trans-Lunar Coast 5) lunar orbit 6) LM landing 7) lunar lift-off 8) rendezvous and docking with CM/SM 9) Transearth Insertion 10) Transearth Coast 11) Splashdown.
(Andrea K. Dobson-Hockey)

Earth and the Moon at 11:22 P.M. on Friday. At 1:28 P.M. the next day, *Columbia* swept behind the Moon.

July 20, 1969, was a day destined to go down in the annals of humankind. At 1:11 that afternoon, *Eagle* unlatched from *Columbia* with Armstrong and Aldrin aboard. Collins, alone in the CM, started his long vigil while his companions made history.

Eagle slowly descended. The world watched and waited. At 1,000 meters above the surface, the astronauts observed the lunar plain up close for the first time. Armstrong saw that the LM was being taken into what he later described as "a field of boulders."

Bypassing the computer that was taking *Eagle* into this rough terrain, Armstrong took manual control of his spacecraft. He steered away and hovered, looking for a good landing site. With less than 10% of his fuel left, Armstrong continued the descent, piloting the lunar lander to a smooth area while Aldrin read off data: ". . . lights on . . . down two-and-a-half . . . forward, forward . . . forty feet, down, two-and-a-half, kicking up some dust . . . thirty feet, two-and-a-half down . . . faint shadow . . . four forward . . . four forward . . . drifting to the right a little . . . down a half . . .forward . . . contact lights, O.K., engine stop . . . descent autocommand override off, engine arm off. 413 is in."

MISSION CONTROL: "We copy you down, *Eagle*."

ARMSTRONG: "Houston, Tranquility Base here. The *Eagle* has landed."

MC: "Roger, Tranquility. We copy you on the ground. You've got a bunch of guys about to turn blue. We're breathing again. Thanks a lot."

Later that night, Armstrong and Aldrin prepared to leave the LM. On the Earth, 403,625 kilometers away, millions followed in anxious anticipation. First Armstrong backed out of the hatch onto the porch and then started down the ladder. Mission Control watched him by means of a small television camera on the side of the LM:

MC: "Okay, Neil, we can see you coming down the ladder now."

ARMSTRONG: ". . . I just checked getting back up to that first step. It's not even collapsed too far, but it's adequate to get back up. . . ."

MC: "Roger, we copy . . . Buzz, this is Houston, f2 one one-sixtieth second for shadow photography on the sequence camera."

ALDRIN: "O.K."

ARMSTRONG: "I'm at the foot of the ladder. The lem footpads are only depressed in the surface about one or two inches although the surface appears to be very, very fine grained as you get close to it. It's almost like a powder . . . I'm going to step off the lem now . . . That's one small step for man—one giant leap for Mankind." (Figure 7–4).

With these words began a three-year exploration unequalled in the history books: the exploration of the Moon. The time was 10:56 P.M.

Ever so carefully at first, Armstrong began to investigate the landing site. "The surface is fine and powdery," described Armstrong, "I can kick it up loosely with my toe. It does adhere in fine layers like powdered charcoal to the sole and sides of my boots."

(NASA photo)

7–4 A human footprint on the Moon.

On the Earth life seemed to stand still as people everywhere watched. In major airports and bus terminals large display screens showed Armstrong walking about on the Moon. Probably never before had so many people in so many different nations joined in watching one thing together.

On the Moon Armstrong's first activities were to take pictures and contingency samples. Contingency samples were specimens to be acquired immediately so that in case an emergency made it necessary to leave the lunar surface in a hurry, scientists would still get some samples to study.

Then it was Aldrin's turn to come down the ladder. They set up the American flag together (Figure 7–5).

(NASA photo)

7–5 Aldrin prepares to walk on the Moon. This photograph was taken by Armstrong, who was already on the lunar surface.

The astronauts were asked to stand by. Mission Control radioed the men that they had a message on the way, and they waited patiently while Richard Nixon spoke to them from the White House.

The prototype Apollo Lunar Surface Experiment Package (ALSEP) was unloaded next. This solar-powered experiment station would send back its greatest bulk of information to the Earth long after the astronauts had left the Moon.

Then Armstrong and Aldrin gathered more samples. The moonwalkers reported that the dust that seemed to cover everything continually caked on the men's boots and moonsuits. The Apollonauts worked strenuously. Armstrong's heart rate once registered 160 beats a minute.

As 500 million people watched on television, Armstrong and Aldrin continued their lunar exploration, picking up rocks and taking pictures. One of the last tasks was to walk a short distance from the LM and bore holes into the lunar crust to remove core samples.

Just before Aldrin climbed up the ladder into the lunar module, he retrieved a solar wind experiment that had been designed to pick up some of the subatomic solar particles drifting out from the Sun while the astronauts were on the Moon. Before his ascent, Armstrong passed specimen boxes up to Aldrin by using a pulley cable.

Two hours, 14 minutes after placing his foot there, Neil Armstrong stepped off the surface of the Moon.

Before leaving the Moon the next day, the astronauts jettisoned several million dollars worth of equipment out the hatch. It was of no more use. Taking up the equipment's space aboard the LM was the precious 21.7 kilograms of rock and soil.

The LM ascent stage lifted off from the Moon at 1:54 P.M., July 21. Collins greeted his partners once again when *Columbia* and *Eagle* redocked at 5:35 P.M. *Eagle* was released to crash onto the rugged lunar terrain below.

Columbia's engine was fired to hurl Apollo XI back toward the Earth at 12:55 A.M., July 22. "Open up the LRL [Lunar Receiving Laboratory] doors," radioed the men in *Columbia* cheerfully.

"No matter where you travel, it's always nice to get home," said Armstrong as Apollo XI passed from the influence of lunar gravity to that of the Earth.

Apollo XI splashed down in the Pacific at 12:15 P.M., July 24, 1,468 kilometers southwest of Honolulu. Men had been to the Moon and had returned safely.

Immediately, frogmen were dispatched to disinfect the astronauts for possible lunar organisms harmful to human life. One hour, eight minutes after splashdown, the crew of Apollo XI was on board the aircraft carrier U.S.S. *Hornet.* They were not to receive a typical heroes' welcome, though. Instead, they were whisked into a special quarantine trailer and treated as though they had the plague (Figure 7–6). Indeed, no one knew if the astronauts had brought back with them a lunar disease.

First examinations showed no signs of lunar infection, but nonetheless the men were subjected to 21 days of confinement in the fail-safe LRL. Extensive medical tests were run. These went so far as to inject lunar dust into white mice to incubate potential moon germs. Finally, with no sign of biological contamination whatsoever, the astronauts were released from quarantine.

The human race, as part of its history of exploring new lands, had come to another. People had gone to the

(NASA photo)

7–6 The Apollo XI crew, in quarantine, are greeted by the President on board the recovery ship. *From left to right in the window:* Armstrong, Collins, and Aldrin.

Moon, not as conquerors but as observers striving to understand its hidden knowledge. We could never be bound to one world again.

APOLLO XII

The goal of landing a man on the Moon had been achieved, but Project Apollo went on. Its purpose now was to continue to explore the Moon in greater detail. The Apollo XII landing site was to be near the lunar equator (in Oceanus Procellarum), 1,598 kilometers away from the Apollo XI site. It would be a more rugged area and would provide a second opinion about the Moon. Although more difficult as a landing site, it was expected to yield more information about the Moon than did the first one.

LM No. 6 was to be called *Intrepid,* the CM, *Yankee Clipper.* These obviously Navy names were chosen by a Navy crew: Charles "Pete" Conrad, Alan Bean, and Richard Gordon—all naval officers.

November 14, 1969, was a dark and stormy day. An intermittent drizzle fell. However, it did not seem likely that the huge Saturn V could be damaged by wind or rain as small rockets might be, so as storm clouds thundered, the countdown progressed without stop.

Briefly, a gap in the clouds opened. Apollo XII lifted off (Figure 7–7). It was almost immediately enveloped by clouds. Suddenly, only seconds later, all power went out aboard the spacecraft. Momentarily, backup generators were called upon to supply vital electricity. Conrad reported anxiously from *Yankee Clipper:* "I don't know what happened. I'm not sure we didn't get hit by lightning."

"We've had a couple of cardiac arrests down here, Pete," replied Mission Control.

"We didn't have time up here."

Later the general consensus was that the metallic Saturn had acted as a ground for a static discharge as the rocket rose into the clouds, which in turn, had shorted out the spacecraft. It was decided that the accident had had no

(NASA photo)

7–7 Lift-off for Apollo XII.

effect on the overall mission, and Apollo XII traveled on, lifting majestically high into the sky where no weather could interfere with it.

Yankee Clipper fired into translunar coast and docked with *Intrepid.*

On November 17, Apollo XII slid silently behind the Moon. Its crew initiated a burn to secure an orbit. Once over the Nearside, the astronauts described the lunar terrain as "sort of a very light concrete."

Bean elaborated: "In fact, if I wanted to look at something that I thought was about the same color as the Moon, I'd go out and look at my driveway." He later continued his description as *Intrepid* flew over the Mare Fecunditatis: "Just a slightly darker gray, looks like the beach sand down at Galveston whenever it's wet."

"Okay," responded Houston, "we had a team of geologists checking your driveway. We'll send them to Galveston now."

"I can't believe it!," "Amazing!," and "Fantastic!" were the excited adjectives transmitted from *Intrepid* as Conrad and Bean descended toward what had been dubbed "Pete's Parking Lot." Touchdown. High above, on his sixteenth orbit, Dick Gordon spotted, through high magnification, the lunar module *Intrepid* sitting on the mare.

Conrad and Bean became the third and fourth men to walk on the Moon. "That may have been a small one for Neil, but that's a long one for me," paraphrased Conrad as he stepped off the ladder. Years later Conrad would star in a television commercial with the theme: "You may not know my face, but . . ."

Unfortunately, viewers on Earth were robbed of the opportunity to watch the astronauts on television. The $78,833 camera, intended to be the first to send color pictures from the Moon to the Earth, was accidentally pointed toward the Sun and permanently damaged.

Conrad had fought to minimize the ceremonies on the EVA and to get directly to work. So it was. The American flag was quickly raised, and the moonwalkers went to their tasks.

First, the ALSEP station was constructed. It was nuclear powered and would serve to verify its predecessor's results. It included new experiments: an ionosphere detector, an atmosphere detector, and a magnetometer.

After that, containers of rock specimens were filled and labeled. Bean drove a core tube into the lunar surface by hitting it with a hammer. "That's skilled craftsmanship," he said.

The next day, Conrad and Bean left the lunar lander for a second EVA.

In April 1967, Surveyor III had soft-landed in Oceanus Procellarum, and now, 2½ years later, parts of it were going home. Intrepid touched down only 200 meters from it.

One of the goals of Apollo XII was to retrieve some of Surveyor III in order to study the effect of the harsh lunar environment over an extended period. Conrad and Bean removed some tubing, cable, and a TV camera from the Surveyor and took them back to *Intrepid.* (Figure 7–8).

Intrepid lifted off after 31½ hours on the Moon. Subsequent to docking with *Yankee Clipper,* Conrad jettisoned *Intrepid* to crash into the Moon to test the seismometer he and Bean had put there the day before.

Yankee Clipper fired out of lunar orbit. It returned to the Earth in weather the opposite of that in which it left. The Pacific was calm and sunny. Apollo XII landed only 7.3 kilometers from the carrier *Hornet.*

On December 11, the crew was released from quarantine and welcomed in Houston. Meanwhile, the LRL reported that the Apollo XII moonrocks were bigger and better than those from Apollo XI.

With a successful Apollo XII, American space exploration seemed to have attained a certain infallibility. The textbook mission of astronauts Conrad, Bean, and Gordon seemed proof of this. It was as if nothing could go wrong. So it had seemed—up until then.

(NASA photo)

7–8 One of the Apollo XII astronauts examines Surveyor III. The Lunar Module is in the background.

APOLLO XIV

"O.K., Houston, we've had a problem here," radioed astronaut Jack Swigert to Mission Control when Apollo XIII was three-quarters of the way to the Moon. This was the first understated announcement of an explosion aboard the Apollo SM that left Swigert, Fred Haise, and mission Commander James Lovell with dwindling supplies of air, water, electricity, and fuel (Figure 7–9). The Apollo XIII landing was cancelled. The safe return of the crew was the result of their own performance and the ingenuity of personnel on the ground who were able to marshal available resources to maintain the men until they could get back to the Earth. The Apollo XIII explosion was not the first serious accident in space, but it was the first in which the crew returned alive.

(NASA photo)

7–9 The damaged Apollo XIII Service Module just after it was jettisoned. The missing panel exposes the area where the explosion occurred.

After a ten-month delay, Project Apollo was up and running again. The final consensus on the Apollo XIII accident was that it was caused by the explosion of an oxygen tank. Modifications in the Apollo spacecraft were made accordingly.

On Apollo XIV rode the fate of the Apollo program, which was picking itself up after its first mission failure in space. Indeed, if success did not come to Apollo XIV, serious questions would be raised by the government and the public (as some already had been) as to the program's importance, cost, and risk.

Commanding Apollo XIV would be Alan Shepard, the first American in space. Shepard's companions on the mission would be Edgar Mitchell and Stuart Roosa, both scheduled for their first flights.

Launch day was January 31, 1971. After a 40-minute hold in the count because of low clouds at the launch site (the lesson of Apollo XII was well remembered!), Apollo XIV was ready. The Apollo XIV Command Module, *Kitty Hawk,* was carried into Earth orbit. The crew had chosen a name for the CM exemplifying aviation history. Kitty Hawk, North Carolina, was the site of the first heavier-than-air flight by Wilbur and Orville Wright less than 70 years earlier. *Antares,* the lunar module, was named for a star used for navigation.

"We'll give it a good ride," reported Shepard.

"Everybody's in great shape," added Mitchell, "having a ball!"

Soon the S4B pushed Apollo XIV out of orbit and toward its goal. That goal was a spot in the Fra Mauro formation, the site where Lovell and Haise had intended to land. On the way to the Moon a 10-second burn of the main engine made up for the time lost in the 40-minute launch delay.

The expended S4B, which had passed the Apollo on its way to the Moon, crashed into the lunar surface and triggered the seismometer there as Apollo XIV swung into orbit. From here, *Antares* separated from the CM with Shepard and Mitchell on board.

On the way down, the *Antares* computer malfunctioned and registered an abort; Shepard bypassed it and steered the lunar lander manually to its target. Touchdown was on February 5.

The same day the two astronauts prepared for their first lunar excursion. "It's a beautiful day in the land of Fra Mauro," announced Shepard as he stepped onto the lunar surface. He and Mitchell planted the 30-by-48-inch American flag. They also placed on the Moon 25 miniature U.S. flags, 50 state flags, and United Nations flags from every member nation.

The next task for the Apollonauts was the deployment of the ALSEP station. It was a more advanced replica of those stations established by other Apollo moonwalkers. Shepard's and Mitchell's activities were relayed to the Earth by a portable television camera, the first to broadcast in color (Figure 7–10).

On the second EVA, the following day, the men headed out to explore. With them they pulled a unique two-wheeled cart that had been designed to increase the Moon explorers' efficiency and manueverability (Figure 7–11). It held various tools, cameras, and collection containers for rock samples. The moonwalkers' principal objective on the EVA was code-named Cone Crater. Taxed, the men fell behind on their schedule. The ascent up the crater walls proved to be more difficult than expected. Shepard fell to his knees once trying to climb the slope to the crater mouth, and his heart rate reached 150 beats a minute. Finally, Shepard told Mitchell, "I don't think we have time to go up there."

"Aw, gee whiz," replied Mitchell, "let's give it a whirl." However, because of lack of time and strength, the stop was cancelled. Back at the LM, the astronauts loaded their samples and prepared to re-enter *Antares* for the last time.

Before climbing up the ladder, Shepard picked out a long-handled device from the mooncart. He fastened an

(NASA photo)

7–10 Mitchell walks from the ALSEP deployment site back to the Lunar Module. The shadow of the photographer, Shepard, is in the foreground.

oblong head onto the rod and dropped two white balls onto the lunar surface. Addressing the TV audience, Shepard told watchers on Earth that he was about to try his hand at golfing on the Moon. "I'm trying a sand shot," he said. Without delay, he lifted the tool, which now did indeed resemble a golf club, and swung at one of the balls. Ludicrously slowly in the lunar gravity, Shepard twisted and made contact with the ball. In a cloud of moondust, it gradually rose and flew out of view.

"Looked like a slice to me, Al," responded Mission Control.

"It goes miles and miles!" Shepard said. Again he hit a ball with the same effect, becoming the first lunar golfer!

(NASA photo)

7–11 Shepard puts together the Modularized Equipment Transporter.

Back in *Antares,* Shepard and Mitchell made ready to rejoin Roosa in *Kitty Hawk.* "What a liftoff!" exclaimed Shepard when the two astronauts rose above the lunar surface. After docking with the CM, the crew orbited the Moon for another day. They rocketed toward home the following morning.

The Apollo XIV astronauts began briefing the *Apollo XV* crew as one mission grew out of the knowledge attained in another. Scientists eagerly opened the lunar samples brought back from the Moon by Apollo XIV and called the mission the most scientifically rewarding yet.

Apollo XV's destination was the Hadley's Rille formation and the surrounding Apennine Mountains. The mission was to be commanded by Apollo IX astronaut, David Scott. His crewmates, Alfred Worden and James Irwin, would complete the Apollo XV team—an all Air Force crew. For this reason it was not surprising that they named their lunar lander *Falcon,* the symbol of the Air Force Academy.

The unique part of the Apollo XV mission was to be a remarkable device in which Scott and Irwin would travel 28 kilometers over the lunar surface, at times being up to eight kilometers away from the Lunar Module. It was the Lunar Rover (LRV). The LRV was a 210-kilogram, collapsible, motorized vehicle that was packed into the side of the LM. When unfolded, the Rover resembled a miniature dune buggy. This dune buggy, however, cost $8 million. It was equipped with an advanced gyroscopic navigation system and an on-board digital computer, instruments that could be used to locate the LM from anywhere on the lunar surface. The vehicle was steered by the driver by using a control stick located in front of the seats. It would carry all the equipment and devices the astronauts would need during their EVA's and would transport rock samples back to the LM from distant sites that the men would visit.

Mounted on the Rover was a self-supported television camera that would beam pictures of the astronauts at work to the Earth. A large antenna atop the mooncar would broadcast the signals. In this way, viewers on the Earth would be able to travel along with the moonriders as they visited different locations near their landing site, far beyond the range of previous crews.

July 26 began partly cloudy, but a large crowd was still on hand, lining the Florida beaches to watch the launch of Apollo XV. The Apollo XV command module was named *Endeavor* after the ship sailed by Captain Cook when he explored the South Pacific in the 1760s. The trip out was uneventful after the Apollo CM docked with the LM and started its course for the Moon. It was July 30 when *Falcon* started its descent to the landing site.

What of the landing site? The Hadley's Rille region was supposed to be the perfect landing target. It had in a very small area five quite different land formations for the astronauts to investigate.

Primarily, there was the sinuous rille itself, a half-kilometer-deep chasm. The moonwalkers hoped to get as close as possible to the rille to obtain samples from it. Also at the landing site were the lunar Apennine Mountains, higher than the mountains with the same name on the Earth, having peaks that rise over 4,000 meters from the lunar surface. Both impact and perhaps volcanic craters in the area were to be compared and contrasted with each other. Last, a flat mare was at the base of the Apennines. It would be used as a control sample for comparison with other maria sampled by previous Apollo missions.

The lunar module landed. "O.K., Houston, *Falcon* is on the plain at Hadley," reported Scott. Shortly after landing, Scott opened the docking hatch atop the LM and stood up through it. He observed the features surrounding the landing site to get a better idea of what the two astronauts could expect to encounter the next day when they explored the lunar valley. "Oh, boy, what a view!" exclaimed Scott.

On July 31, Scott and Irwin began the first of three EVAs, one that was to last a record seven hours. Their initial chore was to set up equipment and unpack the Lunar Rover (Figure 7–12). Because they were to be the first to test it, the two had been given the nickname "The Rover Boys."

The astronauts started out on a short ride aboard the Lunar Rover toward the nearby Apennine Mountains. While Scott drove, Irwin looked out for craters and bumps that might damage the mooncar. When they stopped, their actions there were beamed to the Earth by the color television camera on the Rover.

Next on the agenda was the deployment of the ALSEP. Costing $26 million, it featured the familiar experiments plus a spectrometer and a gravity detector. It completed the trio of nuclear-powered ALSEPs started by

(NASA photo)

7–12 Irwin works beside the Lunar Rover. Mt. Hadley looms in the background.

Apollo XII and Apollo XIV. Only Apollo XI's solar-powered station had ceased to operate.

The following day, the men began a second EVA. They traveled south in the Rover to a group of impact craters and rock formations. At one point Scott discovered a crystal-filled rock of the type scientists on the Earth had expected to be one of the oldest to be found on the Moon. "I think we've found what we're looking for!" Scott excitedly told Mission Control. Back at *Falcon,* Scott and Irwin deployed a heat probe.

The last excursion began the next morning. It was to be the most important. The Rover Boys rode their mooncar in the opposite direction from which they had traveled

(NASA photo)

7–13 Scott operates a 70mm camera. The LRV is behind him. The Apennine Mountains at the horizon are 17.5 kilometers away.

the day before (Figure 7–13). They visited some potentially volcanic craters and Hadley's Rille itself. At the rille, the astronauts were able to climb down part way into the fissure, finding it not as difficult as they had expected. However, it was obvious to those watching from the Earth that the astronauts were working hard. Scott fell to his hands and knees several times trying to obtain rock samples. "This time I'll look and make sure I don't fall over some silly rock," he said disgustedly.

At the close of the EVA, Scott stood before the camera. He removed a tool from the mooncar and a small white quill from a pocket on his moonsuit. He began by saying, "In my left hand I have a feather, in my right hand a hammer . . ." He then explained that nearly 400 years before, Galileo had stated that any two objects would fall at the same speed in the absence of air. "Where better to confirm his findings than on the Moon?" Scott queried. He released both objects from his hands. They slowly fell at exactly the same rate. "That proves Mr. Galileo was correct in his findings," explained Scott.

Scott parked the Lunar Rover permanently several hundred feet from *Falcon.* From here the mooncar would record the ascent of the astronauts from the lunar surface for the first time. On previous missions the moon camera had gotten its power from the LM and was left dead at the moment of liftoff. The Rover camera, receiving its power from the LRV itself, would continue to broadcast. When *Falcon* left the Moon, it seemed to jump off its descent stage and out of view of the Rover camera. Two-and-a-half hours later the three astronauts were once again together in the commandship, and there was time for rest.

Worden had been busy with scientific work of his own 110 kilometers above the Moon, while his comrades explored Hadley's Rille. One of his objectives was to spot and map colored soil deposits, using an assortment of cameras and lenses in the service module. The pictures and information he obtained were stored in cassettes contained in one of the SM bays. After leaving the Moon, Worden would have to go outside the spacecraft and get the cassettes, the next major phase of the Apollo XV mission.

Before saying goodbye to the Moon, the astronauts ejected a small 38-kilogram **subsatellite** from a bay on the SM. This lunar satellite would continue to send back information about the Moon long after the departure of Apollo XV.

After Scott and Irwin had been in the spotlight for three days, it was now Worden's turn. Worden donned his spacesuit and crawled out of *Endeavor* for his 38-minute EVA. He worked his way backward to the service module three times to retrieve his goal, the film cassettes. Worden completed the first deep-spacewalk (Figure 7–14).

Following a normal return to Earth, re-entry, and splashdown in the Pacific, the Apollo XV crew was greeted and congratulated. The quarantine that had been part of previous lunar homecomings had been abandoned. No evidence of biological contamination of any kind had been

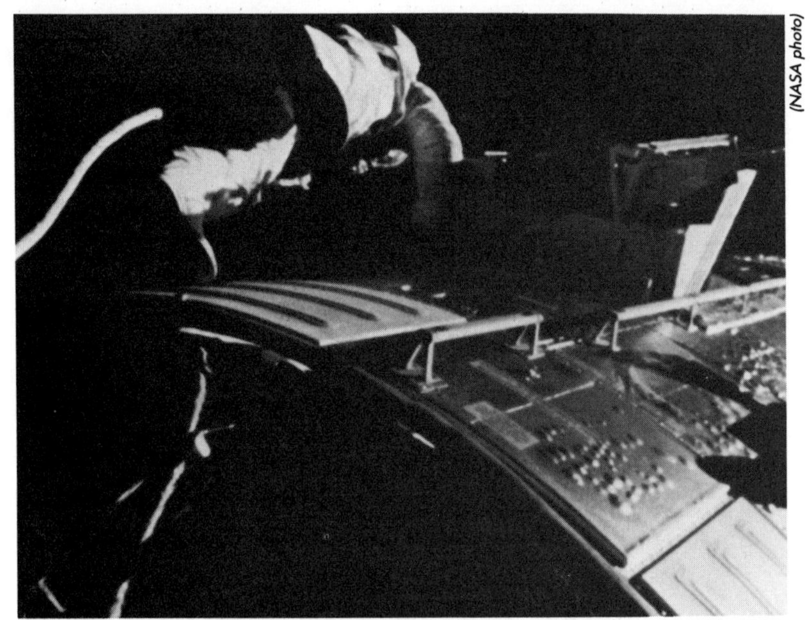
(NASA photo)

7–14 Worden works outside the Apollo XV Service Module.

found. Evidently the Moon was the sterile lifeless world it had been expected to be.

APOLLO XVI

On April 16, 1972, another Apollo spacecraft was launched toward the Moon. Though public attention was at an all-time low, this was an important mission. Apollo XVI's destination was the area surrounding the crater, Descartes. It would be the only actual highlands landing, in a valley below high peaks.

One major change was with the Lunar Rover. A remote-control device had been installed to allow the camera angle to be controlled from the Earth at Mission Control. This would give those watching a better view. The camera would follow the astronauts as they moved about on their tasks.

Aboard Apollo XVI was John Young, a veteran of three previous spaceflights. Occupying the seats next to him were Charles Duke and Thomas Mattingly. The lunar module for the Apollo XVI mission had been given the name

Orion, after a prominent constellation. The Apollo XVI command module was named *Caspar* (after the cartoon ghost).

Apollo XVI achieved lunar orbit on April 19. Forty minutes later, seismometers on the Moon recorded the impact of the Apollo XVI S4B. "Hello, Houston. Sweet Sixteen has arrived," reported Young.

The next day Young and Duke separated *Orion* from *Caspar.* Lunar touchdown came that afternoon. The Apollo XVI landing was calculated to be within 230 meters of its targeted site.

The next day, the two men got into their moonsuits and climbed down the ladder onto the lunar surface. "Here you are, mysterious and unknown Descartes highland plains, Apollo XVI is gonna change your image," said Young as he made his way down from the LM.

While setting up the Apollo XVI ALSEP, Young tripped over a power cable, ruining a $1.2 million experiment. "It broke right at the connector. I didn't even know it. Oh, rats, I'm sorry, Charlie," he apologized. (Later, Duke accidentally dumped a load of equipment he was carrying when he stumbled.)

After assembling the ALSEP, the LM pilot exclaimed, "I'm going out for the Olympics. I just swung the bar on the ALSEP package about 200 meters. Look at that thing go!"

The men found prime rock samples wherever they stopped. On their first moonride, they examined the craters code-named Flag and Spook. "This is so super. Look at all the little goodies!" yelled Duke.

During all the moonwalks, the astronauts' account of what they saw was filled with "wows," "supers," and "fantastics." "Even the craters have craters!" exclaimed Young.

Once the senior astronaut noticed that the taller Duke was gaining more ground than he and paraphrased, "One small step for Charlie is one giant leap for me."

Several times the moon explorers bumped into each

other. Once Young commented, "By golly, we did it again. I never thought we'd run into each other on the Moon!"

When off on the Lunar Rover, the two pilots were right at home. "This is the only way to go," said Duke as Young drove. Later he added, "I feel safe in this thing. Open it up a little bit." (Figure 7–15). After a full day of working, the Apollo adventurers had become the most talkative and enthusiastic team by far to explore the Moon.

The second EVA began on the morning of April 22.

7–15 Looking back at the parked Lunar Rover from the Apollo XVI ALSEP site. The Heat Flow Experiment is in the lower right hand corner of the photograph. Between the Rover and this experiment is the Apollo Lunar Surface Drill. Notice the trail of footprints.

(NASA photo)

"This is gonna be a good day, Charlie," said Young after he stepped off the footpad. This time the astronauts' objectives were to the south. One of them was Cinco Crater, 4 kilometers from *Orion*. The men drove nearly a kilometer up the rim without encountering the crystalline rocks that geologists had expected them to find. While scientists on the Earth puzzled, the moonwalkers drove on to other stops. Finally Duke spotted a rock: "Look at that rock over there. If that's not crystalline rock, I'll depressurize right here!"

As Young and Duke worked, the ruggedness of the terrain became evident. Several times the men tripped or fell to the ground. Tired, they returned to *Orion* for a good night's sleep. EVA No. 3 started the following day. This time they traveled north to Smoky Mountain and North Ray Crater. On this trip the explorers found deep craters and boulders the size of houses (Figure 7–16). Still, the astronauts were able to keep up an average speed of more than 18 kilometers per hour over the tricky landscape.

Young and Duke ended the five-hour, 40-minute excursion with what they called the "Lunar Olympics," which consisted of their leaping and tumbling in the low lunar gravity.

The Lunar Rover, parked in front of the LM, broadcast pictures of *Orion* as it rose silently to meet *Caspar*. The remote-control camera followed the spacecraft until it was but a mere dot in the dark lunar sky. Nine more automated broadcasts were scheduled from the Moon in later days before the Rover power supply went dead.

While still in orbit, the crew jettisoned another subsatellite to circle the Moon. *Caspar* fired out of lunar orbit a day later. "We really haven't even begun to scratch the surface of the complexity of this story," mused Young as he watched the receding Moon.

On April 25 Mattingly duplicated the spacewalk of Al Worden when he left *Caspar* to retrieve film cassettes he had shot while his companions were on the Moon. At the end of his 24-foot safety line Mattingly shouted, "We'll play ride 'em cowboy! . . . Oh, just neat!"

(NASA photo)

7–16 Duke leans his rake against a boulder.

Caspar landed a little over a kilometer from the carrier *Ticonderoga,* the closest splashdown yet, on April 28. Forty minutes later—a recovery record—Young, Duke, and Mattingly were on the carrier deck. "I never thought a group of all males could look so good," Mattingly told the sailors assembled on the deck of the ship to greet the astronauts.

In addition to everything else, the crew of Apollo XVI proved that people can actually enjoy space.

APOLLO XVII

Eugene Cernan was an alumnus of two spaceflights. His most recent, Apollo X, had taken him within 15 kilometers of the lunar surface. This time Apollo XVII would take him the rest of the way. Flying with him would be Ronald Evans and Harrison "Jack" Schmitt. Schmitt, an astrogeologist, hoped to give the most thorough account yet of the Moon's construction.

The Apollo XVII command module was named *America,* the spirit of the United States going with the astronauts on their journey. The lunar module was called *Challenger,* the inspiration to venture into space.

Apollo XVII was to begin at night. On the evening of December 7, 1972, the Saturn V carrying it flooded the pad and surrounding area with a bright red glow. The booster flame resembled a roman candle. Awed spectators watched it until it was but a point of light, another star in the sky. It was a fitting climax to the spectacular launches of Apollo. "Good show, babe!" yelled Cernan.

The astronauts were heading for a place on the Moon called Taurus–Littrow, named for the Taurus mountains and the crater Littrow in the Mare Serenitatis. Cernan would steer *Challenger* to a landing point inside a 2,000-meter box canyon surrounded by tall mountains. It was the last opportunity for lunar exploration for a long time to come.

Apollo XVII slipped into a 60-mile orbit around the Moon. "Oh . . . we're breathing so hard, the windows are fogging up on the inside," reported the enthusiastic Evans as the crew peered out onto the lunar landscape below. Almost immediately Schmitt started to give professional reports on the lunar terrain, using many scientific terms that sent NASA technicians to their dictionaries.

The crew set about preparing for lunar exploration. The LM was readied, and a bay door was jettisoned from the SM to reveal cameras and data recorders. *Challenger* left *America* and slowly sank to the lunar surface below. Finally Cernan reported, "The *Challenger* has landed. We is here man, we is here."

Later that day, the Apollonauts stepped out onto the Moon for a four-hour EVA. As Schmitt crawled out the hatch, Cernan called up to him: "Don't lock it . . . lose the key and we're in trouble."

"We'd like to dedicate these first steps of Apollo XVII to all those who made it possible," said Cernan a little bit later.

On the lunar surface, Schmitt described the landing site as "a geological paradise."

First the men established the ALSEP and unpacked

(NASA photo)

7–17 Cernan test-drives the LRV. The communications equipment and geology tools will be put on board later.

LRV No. 3 (Figure 7–17). While setting up instruments, Schmitt sang, ". . . We're off to see the wizard . . ." The two astronauts also set up the U.S. flag. This particular standard had been hanging in the Mission Control room in Houston ever since the Apollo XI flight. The men then paused for pictures in front of the flag.

The astranauts' last major job was to bore a core sample out of the hard lunar crust. This was found to be extremely difficult. The exertion left the men almost exhausted. Even though the work was strenuous, the moonwalkers happily carried out their tasks. Cernan broke into song, singing, "Oh bury me not on the lone prairie . . ."

On December 12, the men set out on a second EVA. The day before the moonriders had damaged a fender on the LRV. Without its working properly, the Rover wheels threw dust and dirt over the astronauts as they drove. The men made a hasty repair job using maps fastened together to form a makeshift fender, which worked satisfactorily throughout the remaining rides.

The biggest discovery of the second EVA occurred suddenly when Cernan abruptly shouted, "That's orange! . . . That's incredible, it's really orange out here, Houston." What Cernan had found was a patch of soil that was orange even to the TV camera. He and Schmitt quickly took a large sample of it.

7–18 Splashdown for Apollo XVII, southeast of American Samoa. This photograph was taken from a recovery helicopter.

(NASA photo)

After a successful climaxing excursion, the Apollonauts returned to *Challenger.* At the LM, they held a ceremony marking the last probable exploration of the Moon in the decade. Cernan and Schmitt unveiled a plaque attached to a landing leg of the lunar module. It read: "Here Man completed his first exploration of the Moon, December 1972 A.D. May the spirit of peace in which we came be reflected in the lives of all Mankind." ". . . I believe history will record that America's challenge of today has forged Man's destiny of tomorrow," stated Cernan.

Before climbing up the ladder, Schmitt hurled his geologist's hammer over the lunar plain. Eugene Cernan, the last Apollo astronaut on the Moon, marked the end of an all-too-short era at 12:35 A.M., December 14, 1972.

Back with Evans and *America,* the Apollo XVII crew said farewell to the Moon and set course for the Earth. Sunday morning the crew was awakened by *Home for the Holidays* relayed from Houston. Evans carried out another deep-spacewalk to bring film cassettes in the Service Module back to the Command Module. Floating 300,000 kilometers from the Earth, Evans acknowledged the camera pointing at him and greeted his wife and children.

America ended its journey on December 19 (Figure 7–18). Apollo XVII was the end of the program. Although Project Apollo had left many questions unanswered, it was still the first small step into a universe in which we would no longer be complete strangers. "This is just the end of the beginning of a golden age of space flight," said Cernan.

8 THE MOON UP CLOSE

THE LUNAR SURFACE

The first discovery about the Moon's surface made upon landing the Luna and Surveyor spacecraft was that they stayed there. They didn't sink. Even after the Ranger flights, it was still not known for sure whether the lunar surface would support a vehicle or, for that matter, a person standing on it. The maria were chosen as the first landing sites because of the apparent lack of obstacles there, but what were the maria made of?

In 1955 Thomas Gold proposed that the lunar maria might be covered with layers of moondust eroded away from the highlands and transported to the lower basins by electrostatic forces to accumulate there. Gold suggested that this dust might be the density of water and tens of meters deep. A spacecraft landing in this material wouldn't land at all; it would sink to the bottom of a sea of powder. This potential catastrophe (and ways of dealing with it) was fictionally illustrated in Arthur C. Clarke's 1961 best seller, *A Fall of Moondust.*

Even if the lunar surface would support astronauts and equipment, it still might be treacherous, it was thought. It is difficult today to remember the extent of our unfamilarity with the lunar surface before we landed on it. As late as 1960 it was still suggested that the maria might be the sediment of ancient seas! At least you could walk on a seabed. Some said that the maria might be filled with a frothy slag or sharp jagged stones, both difficult surfaces to tread over. For this reason, a major objective of the Surveyor program was to study the lunar surface and see if it was safe for people. The surface was found to be solid and covered with a firm "soil" (fine particles clumped together in the vacuum of the lunar environment). Scattered over the soil were moonrocks of every size (see Figure 8–1). The Apollo astronauts found that the lunar soil stuck to their instruments. The trenches they dug did not cave in readily, and

8–1 The first picture taken by Surveyor I shows the footpad *(center)* unburied and firmly planted on the lunar surface.

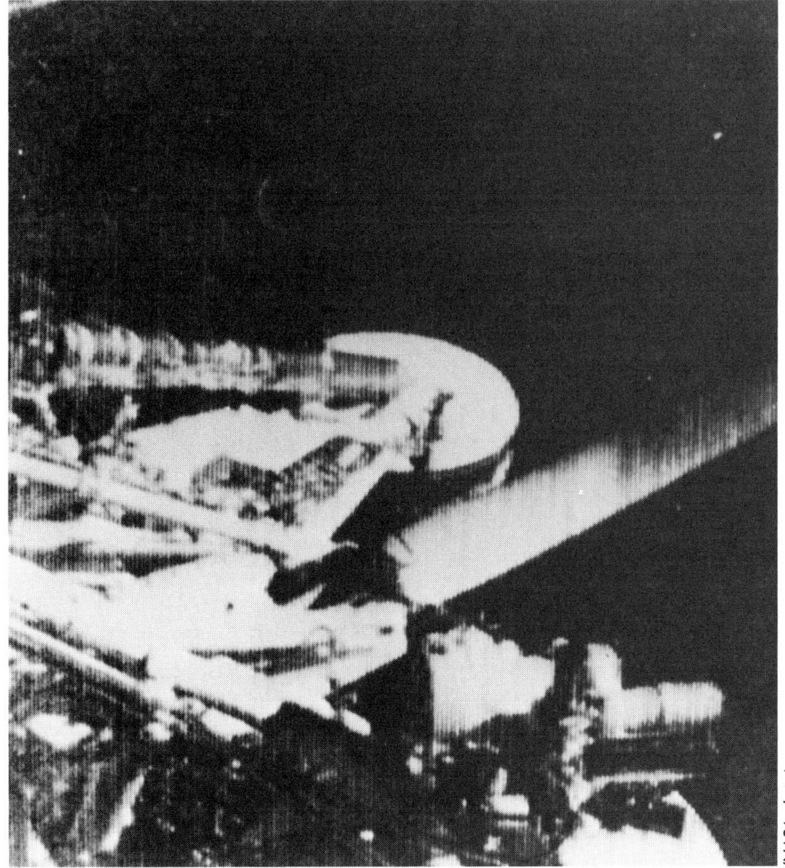

(NASA photo)

their footprints did not slump. In fact, the soil behaved very much like wet beach sand.

The lunar soil is really a layer of loose particles called the **regolith.** These particles, in a layer several meters thick in the maria, are the product of repeated pulverization of the lunar surface by large and small impacts over the eons. The continual churning of the particle layer by these impacts results in the fine, homogeneous layer that the astronauts walked on.

The TV cameras showed us a lunar landscape that appeared to be serene and unchanging. One of the benefits of going to the Moon was that geologists were able to study a surface unaltered by traditional processes of erosion such as wind, rain, and life that have totally obliterated the ancient surface of the Earth many times over. The lunar surface does change, though—albeit on a much slower time scale than does that of the Earth.

If a meteoroid weighing 40 million kilograms and traveling at 15 kilometers per second hit the Moon, it would create a crater a kilometer across. It and subsequent secondary and tertiary events could eventually displace 1,000 times this mass in rock and regolith (see Figure 8–2). Such an event is rare, however. It happens only once every 3 million years or so. Smaller bodies are more efficient sources of erosion. For instance, on a given day on the Moon only one object weighing 10 or more grams will strike an area of 70,000 square kilometers. However, on the same day, each square foot of the Moon is hit by at least one micrometeorite. Micrometeorites make up 99% of the 40,000 kilograms of material that rain down on the Moon every twenty-four hours. These projectiles may be no more than 0.001 millimeters in diameter, but they are responsible for a large part of the erosion on the Moon. It has been estimated that micrometeoroids churn over the entire regolith every 40 million years.

More subtle is the effect of **cosmic rays,** subatomic particles from space that continually strike all exposed surfaces unprotected by a blanket of atmosphere. On the Moon,

(NASA photo)

8–2 The interior of a typical mid-sized lunar crater discovered by Apollo XVI astronauts Young and Duke. Crater formation on this scale is infrequent.

cosmic rays cause chemical changes in the soil and damage the crystalline structure of rocks. Their penetration power is limited, however. Only exposed surfaces are affected. Thus, they serve as an indicator of the movement of moonrocks. The astronauts recovered samples that showed cosmic ray damage on the undersides. This meant that these rocks had been turned over some time in their recent history. Likewise, buried rocks that show the effects of cosmic rays must once have lain on the surface.

Movement of moonrocks may be caused violently by the scatter from impacts or less dramatically by the inexorable force of gravity. A boulder teeters for centuries and finally rolls down hill; lunar soil slowly slips down an incline. Such processes are known as **mass wasting** and may happen continuously or be triggered by nearby impacts or tectonic activity.

A less obvious source of change on the moon is

(NASA photo)

8–3 Apollo XVII crew member Schmitt stands next to a huge boulder that has been split in two.

erosion caused by the extreme variations in temperature endured by surface rocks as night becomes day. This diurnal thermal range may span 250°C. Some individual rocks may be heated to as hot as 130°C. in the daytime. This cycle of temperatures causes expansion and contraction in the moonrocks that eventually induces cracking and crumbling (Figure 8–3).

Not included among these sources of lunar erosion are astronauts who pick up rocks, dig holes, set off explosives, and do other things to violate the lunar surface!

The lunar regolith does not conduct heat as well as does rock, as evidenced by infrared pictures taken of the Moon during lunar eclipses. The lunar disk rapidly cools in the Earth's shadow, but certain areas—hot spots 30°C. warmer than their surroundings—remain glowing in the infrared long afterward. Such hot spots are likely to be regions that have a high abundance of surface rocks that soak up heat while they are exposed to the Sun. This heat is reradiated into space in the infrared during the eclipse. The lunar soil does absorb as much heat, however. It is, in fact, a pervasive insulating blanket that wraps the entire surface of the Moon because it is made up of small particles packed together in a vacuum. The effect is much like our terrestial vacuum bottles: A cup of coffee buried in lunar soil would stay warm.

Many of the hot spots on the Moon are in the vicinity of young craters where presumably the regolith has been blown away and has not had time to accumulate again. They are holes in the Moon's insulating blanket.

Looking at longer infrared wavelengths allows us to "see" further into the surface. At one meter in depth there is little temperature fluctuation; it is nearly a constant: − 50°C. day and night.

While the Apollo astronauts examined selected sites on the Moon in detail, selenologists on the Earth compiled the pictures taken by both manned and unmanned orbital spacecraft for a global view of the Moon. Their aim was to compose a geologic map of the Moon that would delineate different rock formations to be found there. Using **stratigraphy,** they were able to assign relative ages to these formations. Stratigraphy refers to the order in which layers of rock are laid down. For instance, smooth dark areas that interrupt otherwise rough terrain can be assumed to have covered portions of older terrain and must therefore be younger than those they overlap. Craters in the smooth material ought to be younger still. Likewise, a fault that displaces one side of a feature must be younger than that feature, and a crater that is superimposed over the rim of another must be the younger of the two.

Stratigraphy necessitates that some part of a rock stratum be exposed. On the ground, craters serve as an excavation tool for examining buried strata. The bowl-shaped depressions may have cut through one or more different strata. Thus, rock samples collected from near the bottom of a crater can be of a different age and composition from those of the rim.

Another tool for assigning ages to terrain is **crater counting.** The idea is that the more craters that can be found in a given area, the longer that region has existed and been exposed to meteoric bombardment. Ranger VII first revealed that this technique breaks down where a surface is saturated—that is, where the crater density is so high that there is no room for any more! The formation of a new crater necessitates the destruction of an old one. Thus, the surface is in a steady state, and the number of craters at any given time remains constant.

In the case of a steady-state surface, on-site drilling could measure the depth of the regolith and determine roughly how long that surface material has been pelted and broken up by impacts. All sampling techniques such as these, though, run the risk of contamination by material that was displaced from some other place on the Moon. For instance, the Apollo XIV and XV expeditions both landed on the ejecta blankets of the distant Mare Imbrium.

Earlier, I suggested that the surface of the Moon might be modified by tectonic events. What evidence is there for such events to have happened? The Moon does not have large-scale, global tectonic features, as do some other planets and satellites. Rather, the ridges and rilles seem to evidence a more local and less violent type of tectonic event. Many of the rilles appear to be true crevasses. Some are likely to be **graben faults.** Warping may have created arcuate rilles. The Straight Wall is a classic example of a lunar fault.

We return now to the matter of volcanism on the Moon. Once it was decided that impacts account for most of the craters on the Moon, the question shifted from the origin of the craters to whether there had been any volcanic activity on the Moon at all and if so when. We have already spoken of certain features that may be **endogenic** (formed by internal processes) rather than **exogenic** (formed by external processes): dimple craters, halo craters, and certain areas such as the regions surrounding the craters Alphonsus (Ranger IX was specifically targeted for Alphonsus to check out this theory) and Aristarchus. We can add to this list craters lacking in ejecta and certain steeply domed features that suggest a thick lava squeezed out of the Moon's interior and piled up.

The sinuous rilles represent strong evidence for volcanism. Their "head" craters are suspiciously volcanic. These rilles may be the remains of lava tubes flowing away from a volcanic vent. The vaulting roofs of such tubes could become quite broad under the low lunar gravity. Eventually, though, the roofs might collapse and form the exposed rilles that we see.

An explanation for the wrinkle ridges is that they were formed by lava oozing out of long fissures (see Figure 8–4). Less controversial are a number of clearly recognizable

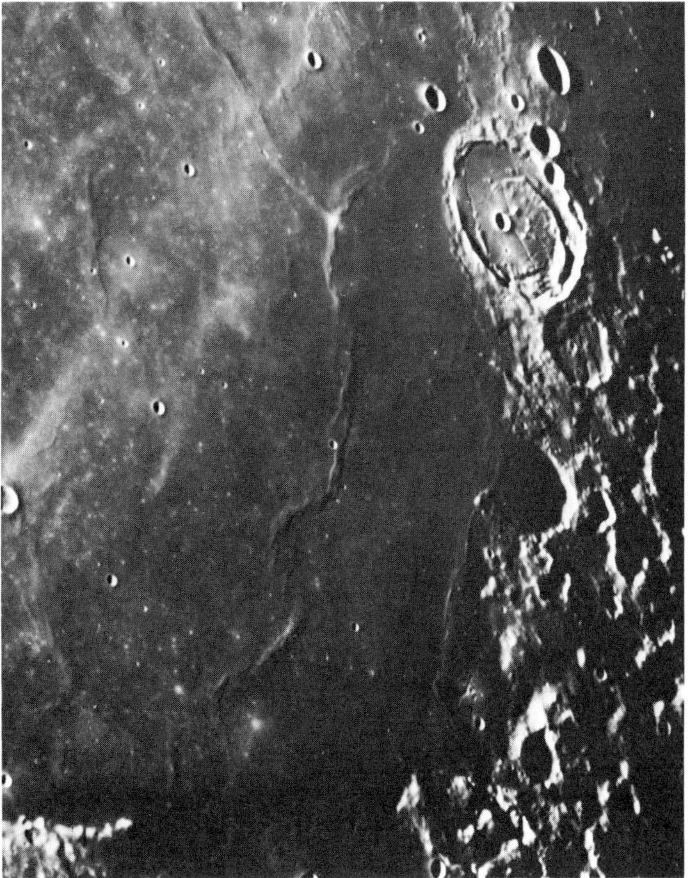

(Lick Observatory Photograph)

8–4 Wrinkle ridges in Mare Serenitatis. The large crater is Posidonius.

lava flows on the Moon. These flows are distinguished by sudden changes in the lunar albedo. Similarly, the insides of some craters appear to have been flooded, and the maria are thought to be the result of lava flooding early in the Moon's history.

After Apollo there is a consensus that the formations on the Moon can be divided into those that are definitely not volcanic, those that are, and a great many that are ambiguous. For instance, Apollo XVI was directed to a region that was thought to be endogenic—the Cayley Formation. Everywhere the astronauts went, though, they picked up samples that appeared to have been deposited by impacts!

It was the last Apollo that brought back what was perhaps the most interesting example of volcanic processes found on the Moon. Color on the otherwise drab Moon is a tipoff of something unusual. For this reason that astronauts and ground geologists alike were interested when orange soil was discovered during Apollo XVII (Figure 8–5). This soil was found to be a fine volcanic glass, a product of **Fire fountaining.** Similar but less spectacular green glass was found in Apollo XV samples.

Somewhat ironically, the Apollo XVII astronauts had been looking for low albedo areas associated with halo craters at their landing site. These were not found owing to the difficulty in identifying them so close up. (John F. Lindsay, *Lunar Stratigraphy and Sedimentology,* Amsterdam: Elsevier Scientific Publishing Co., 1976). A most promising Apollo landing site for finding volcanic features, the Marius Hills region, was never visited because later flights were cancelled (see Chapter 9) (G. Fielder, ed., *Geology and Physics of the Moon,* Amsterdam: Elsevier Scientific Publishing Co., 1971).

The Apollo astronauts encountered no erupting volcanoes or flowing rivers of molten lava on the Moon. The current opinion is that although TLPs may be evidence of outgassing or the release of ash, active volcanism does not

(NASA photo)

8–5 The famous "orange soil" was first spotted by astronaut-geologist Harrison Schmitt.

seem to be present on the Moon now. When did it take place, then?

Stratigraphic and other arguments are used to date lava flows. They tell us, for instance, that the mare basins were filled starting about 3.8 billion years ago. Mare flooding lasted for hundreds of millions of years. This may have been the heyday of lunar volcanism. (We do not know, though, why the basins on the Farside have not been filled.)

The lunar orbiters discovered signs of the upwelling of magma associated with maria. While monitoring these spacecraft as they circled the Moon, engineers discovered alterations in their orbits, alterations that could be accounted for by local changes in gravity produced by abnormal concentrations of mass (**mascons**) beneath certain maria. The biggest of these mascons lie below Maria Imbrium, Serenitatis, Crisium, Nectaris, Aestuum, Humorum, Humboldtianum, Orientale, and Smythii. The gravitational effect of each is roughly proportional to the area of the mare that overlies it. The engineers went on to adjust the orbits of future spacecraft (such as Apollo) to take the mascons into account; the scientists wondered why these abnormalities existed. Early theories said that they represented metallic asteroid-size bodies lying under the maria. However, such bodies would have had to be traveling unreasonably slowly when they implanted themselves in order to remain intact. More recently, it had been suggested that the mascons are plugs of high density **magma** that have risen beneath the basins. If this is so, the emplacement of this material might have taken place late enough in the Moon's history so that the satellite had become rigid and able to support the mascons near the surface. Thus, the formation of the mascons may be just one of the events of the age of lunar volcanism.

Selonologists today place heavy lunar volcanism during a period 3.8 to 2.5 billion years ago. However, some would have it beginning as early as 4.3 billion years ago (nearly the age of the Moon itself) or ending as little as 1 billion years ago. ("A More Active Moon?," *Sky and Telescope,* **66,** December 1983, 501). The exact period depends on the interval during which an internal heat source within the Moon existed to drive volcanic activity to the surface. We take up the subject of this heat source in a subsequent section.

We can now put together a brief history of the Moon's surface. After its formation, it resembled the lunar highlands of today. Its face was battered by a period of heavy meteoric bombardment that lasted for the first seventh of the Moon's lifetime. Large mare basins were formed by the sweeping up by the Moon of other "moonlets" orbiting the Sun, the Earth, or the Moon itself (S. K. Runcorn, "Implications of Lunar Paleomagnetism for the Origin of the Moon," abstract, *Papers Presented to the Conference on the Origin of the*

(Don Wilhelms and Don Davis, USGS)

8–6 The Moon 3.8 billion years ago. Notice the lava just beginning to flow into the large impact basins.

Moon, 1984, p. 10). The impact of the mare-forming bodies raised the lunar mountain ranges and threw debris over most of the rest of the lunar terrain. As bombardment subsided, intermittent lava flooding began (Figure 8–6). In time, the lowlands were covered by mare material originating in the basins; the Moon appeared as it does today except for a few large craters that appeared off and on over the millenia (Figures 8–7 and 8–8). Subsequent erosion altered the Moon's face only slightly. Its major effect was to erase the delicate rays (now taken to be ejecta structures) of all but the youngest craters. So the Moon rested until it was disturbed by the footprints, tire tracks, and high tech litter of Apollo.

(Don Wilhelms and Don Davis, USGS)

8–7 The Moon 3.3 billion years ago. The basins have flooded and the new maria have smoothed out much of the rugged, more ancient terrain.

MOONROCKS

In Houston, Texas, 375 kilograms of rock lie in storage. They are divided into 19,000 individual specimens, each of which is wrapped in Teflon and packed in sturdy containers. The containers in turn are placed in trays stored in cabinets within a large vault. These rocks were brought here at enormous cost. They are the moonrocks.

Their home is the Lunar Receiving Laboratory. (Activities of the LRL are described in detail by Andrew Chaikin in "Pieces of the Sky," *Sky and Telescope,* **63,** April 1982, 344). Here, using quarantine methods developed in germ warfare studies, most of the rocks remain untouched by human hands. They are stored in nitrogen gas (an inert atmosphere in lieu of the hard vacuum of the Moon) with an airlock between them and the outside world. They are han-

(Don Wilhelms and Don Davis, USGS)

8–8 The Moon today. The maria have been scarred by relatively recent impact craters and ray systems. Older craters have been eroded somewhat.

dled through a glove box. A separate glove box exists for samples from each landing site.

Elsewhere in the LRL the soil and core samples are kept. The entire building seems well protected; even so, a representative "insurance" set of samples (54 kilograms) has been carefully separated out and stored at nearby Brooks Air Force Base in case some calamity befalls the LRL. About 8 kilograms of moonrock are now in various museums around the world.

To come in contact with the lunar samples, scientists must dress in sterile coveralls, boots, gloves, and caps—much like surgeons entering the operating room—and then take an air "shower" to blow-off any remaining contaminants they might take in with them.

When a sample is used, the specimen as well as any chips of dust produced are carefully weighed to make sure nothing is lost. When a rock is broken, the pieces are photographed to show how they are related to one another.

Within the LRL, the moonrocks have been subjected to nearly every geochemical test and experiment known. Even so, a significant amount has been left unused for future experiments that haven't been thought up yet. Has all this special handling been worthwhile? What have we learned so far about these precious moonrocks, our first samples of a world beyond our own?

The initial Apollo landing site was on a flat mare. One of the initial successes of Apollo was the confirmation of the nature of the maria. They do indeed consist of lava flows from the interior of the Moon. This type of igneous rock is called **basalt.** The flowing molten basalt was not very viscous when it erupted to the surface. It surfaced at a temperature of about 1,400° C. and was the consistency of motor oil (J. E. Guest and R. Greeley, *Geology on the Moon,* London: Wykeham Publications, Ltd., 1977). Hence, it was able to spill out over the mare basins rapidly. This low viscosity also made it easier to form the crystalline basaltic rock we see today.

We know that these rocks were formed at the surface after the molten lava flowed onto it because of their small grain size. Also, many of the crystalline rocks have holes in them that were formed by gas boiling out of the rock. These holes are called **vesicles** when they are spherical and **vugs** when they are irregular. The boiling occurred at the surface when the pressure on the magma was reduced. The gas bubbled out of the molten rock like carbon dioxide out of an opened can of soda pop.

The pristine nonmare rocks are made up mostly of **anorthositic gabbro** and certain other magnesium-rich components. The lunar basaltic rocks are largely composed of the mineral plagioclase. Plagioclase is a course-grained mineral made up of a white granular aggregate. The number of possible minerals is somewhat restricted because of the absence of water. There are no hydrated minerals on the lunar surface.

Once a mineral forms out of a molten mixture, it either floats or sinks. Thus, the particular minerals that we see in the surface moonrocks are somewhat dependent on their melting temperatures. Plagioclase solidifies at a relatively high temperature. Thus, this mineral formed early in the process of rock formation and rose to the surface.

Two new minerals were discovered on the Moon as soon as the first box of samples was returned to the Earth: Tranquillityte and Armalcolite. The first name comes from the Apollo XI landing site. The second is a contraction of the names of the Apollo XI crew.

Finding basalt in the maria was not surprising; finding a great deal of it in the highlands was. The highland basalt is extemely old and may co-date the dense cratering we see there today.

A peculiar kind of basalt on the Moon is called **KREEP.** This curious acronym stands for potasium (chemical symbol K), rare-earth elements, and phosphorus. KREEP has an unusual amount of these elements in it. It is a contaminant found at all of the Apollo landing sites. It is thought that KREEP comes from deep in the Moon's **crust** and was scattered all over the lunar Nearside by large impacts.

Many of the returned lunar samples are not single rocks. They are **breccias,** fused fragments of other rocks pulverized by repeated impacts. Microbreccias are lithified portions of the regolith. They are sedimentary rocks formed by the heat and pressure generated by additional impacts. Some breccias are broken up and then formed into new breccias. Subsequently, we find breccias made up of pieces of other breccia. Breccias may demonstrate the effects of severe damage. Some show evidence of **shock metamorphosis** caused by nearby later impacts.

An extreme effect of shock is the formation of glass. Molten minerals produced by the heat of impact are thrown upward. Exposed to space, they cool quickly into glass spherules and teardrops before falling to the ground. Sometimes the glasses have flecks of metal such as iron embedded in them. Micrometeorites produce tiny pits on rock faces, and the pits are sometimes lined with glass.

Volcanic fire fountaining can produce glass in a similar way. Twenty-three different kinds of volcanic glass have been identified on the Moon (C. Koeberl, "Volatile Elements In and On Lunar Volcanic Glasses: What Do They Tell Us About Lunar Genesis?" *Papers Presented to the Conference on the Origin of the Moon,* abstract, 1984, p. 23).

No new elements have been discovered on the Moon. None were expected. However, selenologists wanted to know what the relative *abundances* of the elements were on the Moon and how these compare to those on the Earth and other bodies in the solar system.

At first glance, the Earth, Moon, and the rest of the solar system are pretty much the same in this respect. They contain the elements detected or caluclated to be in the Sun in the proportion (with some differences caused by subsequent evolution) that existed in the primordial material out of which all these bodies formed.

On closer examination, though, the Moon resembles the outside layer of the Earth more than either resembles the Sun. A rough analogy for the lunar surface is the terrestrial ocean beds. There are differences, however, that are important in trying to assess how the Earth and the Moon are linked geochemically.

A handicap in determining abundances for the whole Moon (or Earth) is that we can look directly only at the crust. We must try to infer if what holds true for the crust also applies to the interior.

I have already said that most of the light, rare gaseous elements (helium, neon, argon, krypton, and xenon) have long since disappeared from the Moon. Some rare gas atoms are **radiogenic:** They are products of the decay of heavy radioactive elements inside the Moon (see Chapter 5). Some may also be found, though, as **fines** in the lunar regolith and crystalline rocks. The fines have been implanted there by the solar wind and cosmic rays. The solar wind is likely to be responsible for much of the hydrogen, carbon, and a little of the nitrogen on the Moon, too. These elements are all said to be **cosmogenic.**

In comparison with the rest of the solar system, the Moon is depleted of additional **volatile** elements such as sulphur, chlorine, selenium, and bromine. These are elements that are easily vaporized at high temperatures. This depletion is a characteristic that the Moon shares with certain types of meteorities. (There were some volatiles found in the volcanic glasses from the Apollo landing sites.)

On the other hand, the Moon is abundant in the **rare earth** elements relative to the Earth (remember KREEP). These elements have exotic sounding names that most of us are not familiar with here on the Earth: lanthanum, cerium, praseodymium, neodymium, promethium, samarium, europium, gadolinium, terbium, dysprosium, holmium, erbium, thulium, ytterbium, and lutetium.

The Moon also seems to be rich in other **refractory elements.** Unlike the volatiles, these elements can exist as gases only at extremely high temperatures. The highlands are rich in aluminum, calcium, and magnesium. Perhaps more unexpectedly, parts of the Moon have a high titanium content—a fact first discovered by Surveyor V.

The Moon appears to be deficient in certain other metals. One such group is the **siderophilic elements:** sodium, iron, nickel, zinc, germanium, arsenic, selenium, bromine, silver, cadmium, indium, tin, antimony, tellurium, rhenium, osmium, iridium, gold, thallium, and bismuth. “Siderophilic” means “iron loving.” These metals tend to combine with iron to make compounds.

The most important of the siderophiles is iron itself. Iron can be readily found in the lunar basalts as iron oxide. There are more iron compounds in the Moon's crust than in the Earth's. Still, overall, the Moon is decidedly depleted in iron. Much of the lunar surface iron and nickel have been deposited there by meteorites—bodies rich in these metals. (Recall the metal flecks in impact-produced glasses.)

One of the reasons that selenologists were so eager to get rocks from the Moon was the rocks' age. Earth rocks are relatively young. Old Earth rocks no longer exist—erosion and geological activity have seen to that. A 3-billion-year-old Earth rock is ancient. Meteorites may be close to 5 billion years old. In between is a period of geological history that is unaccounted for. This is the interval that the moonrocks span. They fill in the time line of the history of the planets.

Moonrocks are dated by using complex techniques that measure the quantities of certain heavy elements in them. An example is **rubidium–strontium dating.** Rubidium–87 radioactively decays to strontium–87. Strontium–87 is not primordial (i.e., there to begin with). If it is found in a moonrock, it is reasonably certain that that rock has been around long enough for some rubidium–87 present in it to decay into strontium–87. By measuring the ratio of the abundances of these elements in a rock, one can tell how much of the rubidium has decayed and how long the decay process has been at work. The rubidium–strontium technique has achieved geochemical dating to within 100 million years, a small fraction of the age of most lunar minerals.

Radioactive "clocks," such as rubidium–strontium dating, tell us that the ages of the rocks brought back from the Apollo landing sites vary. The Apollo XII site was the youngest, the Apollo XV site the oldest. The ages of the returned moonrocks do begin approximately where samples from the Earth end. Most mare samples range between 3.2 and 3.8 billion years old. The oldest moonrocks come from the highlands and are about 4 billion years old. These rocks are believed to be from some of the earliest lunar crust and may be contemporary with primitive meteorites. In fact, one Apollo XVII rock seems to be nearly as old as the Moon itself (4.6 billion years old)! (Stuart R. Taylor, *Planetary Science: A Lunar Perspective,* Houston: Lunar and Planetary Science Institute, 1982)

INSIDE THE MOON

We can't see most of the Moon. The lunar interior can be studied only indirectly. It is the interior, though, which holds many of the secrets of what the Moon is made of and how it came to be.

We know the mass of the Moon. The volume (obtained by measuring the Moon's radius) is 2.20 10^{25} cm^3. By dividing mass by volume, we ascertain its mean density, 3.34 g/cm^3 compared to 5.52 g/cm^3 for the Earth. This was an early clue that the Moon was geochemically different from the Earth.

Harold Urey (1893–1981) was a pioneer of **planetary science.** (The new field of planetary science does not study one planet or another in isolation but, rather, attempts to find similarities and generalized concepts that can be applied throughout the solar system.) Urey calculated how gravitational compression would affect a body the size of the Moon made out of the same materials in the same relative abundances as the Earth. He arrived at a density of 4.4 gm/cm^3. Thus, on the average, lunar material is less dense than is terrestrial material.

The mean density does not tell us how the Moon's mass is distributed throughout the body. However, because the density of lunar surface rocks that have actually been measured in the laboratory are even less than the average value, we know that the interior rocks must be more dense in order for the whole still to yield the mean result.

We know that on an even finer scale the Moon is not a homogeneous body. By observing how the Moon rotates and how orbiting spacecraft behave near it, it is possible to gain information on the Moon's shape.

Let us return for a moment to libration, a phenomenon introduced in Chapter 4. Remember that libration allows us to see around the edges of the lunar Nearside periodically. There are several kinds of libration; all are owing to the fact that the Moon is relatively close to the Earth. One, optical libration, is simply an effect of perspective caused by the Earth's rotation. Our point of view of the

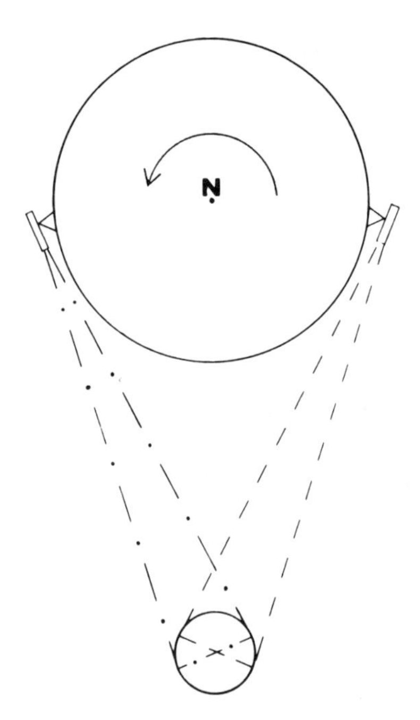

8–9 Optical libration. The face of the Moon observed in the morning is slightly different from that seen in the evening.
(Andrea K. Dobson-Hockey)

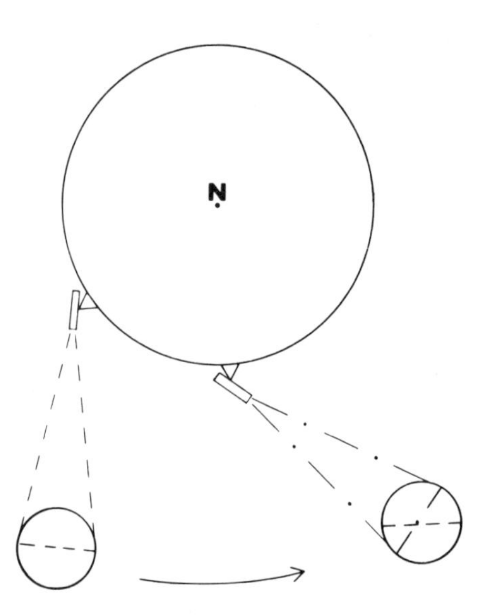

8–10 Physical libration. The Moon is observed from the Earth at two different times. Because of the Moon's rocking motion, a slightly different face is presented on each occasion.
(Andrea K. Dobson-Hockey)

Moon shifts by 12,756 kilometers (the diameter of the Earth) in the course of a night which causes a libration in longitude (Figure 8–9) A libration in latitude is caused by the angle between the Moon's orbital plane and the Earth's equator—sometimes we can "peek under" the Moon.

The libration that we are interested in here can help us learn about the internal structure of the Moon. It is *physical* libration (Figure 8–10). We already know that the Moon travels at different speeds in its orbit (faster at perigee, slower at apogee in accordance with Kepler's Second Law). However, it attempts to rotate on its axis at a constant speed. For this reason, rotation is continually lagging behind or getting ahead of the Moon's orbital motion. Just like the Earth, the Moon has a tidal bulge. Just as the Earth's tidal bulge is caused by the Moon's tugging on the Earth, the Moon's tidal bulge is caused by the Earth's pulling on the Moon. Once this tidal bulge gets out of alignment with the line between the center of the Moon and the center of the Earth, gravity tries to pull it back. The Moon is set to rocking. Other density inhomogeneities in the Moon also serve as "lever arms" and produce the motions that we call libration. By analyzing the rocking motion, it is possible to learn about the Moon's tidal bulge, its shape, and how heavy and light materials are distributed inside the Moon. This analysis has been refined by using the laser **retroreflectors** left behind on the Moon at the ALSEP stations. (The passive retroreflectors are the only ALSEP experiments still being used to collect data.) The uses of the lunar retroreflectors are explained more fully by Derral Mulholland in "How High the Moon: A Decade of Laser Ranging," *Sky and Telescope,* **60,** October 1980, p. 274.

A retroreflector redirects light that falls upon it back in the direction from which it came. A laser produces a coherent beam of light that spreads out very little as it travels. By directing a laser beam through a telescope, the beam can be aimed at a lunar retroreflector. By strobing such a laser and using a telescope to watch the return flash from the Moon, the light's travel time can be measured. Because light travels at a constant velocity, simple arithmetic yields the distance to the retroreflector at that moment. By measuring the time to the billionth of a second, a phenemonal accuracy—within 10 centimeters—can be achieved. The current value for the Moon's radius and the distance to the Moon, for instance, are based on laser ranging.

Although simple in theory, this process is tricky in practice. Not all of the beam hits the retroreflector or intercepts the telescope on its return. The atmosphere disperses and absorbs part of the beam. The returning flash is visible only with the aid of a photomultiplier, which can count only one or two photons returning from the Moon (out of billions sent). Spurious photons are always a problem. Each incoming photon is electronically queried to see if it is an actual return: Did it come in at approximately the right time? Is it the right color? Statistical processes are used to build up an intelligible signal.

By using laser ranging to watch the lunar ALSEP swing back and forth, it is possible to draw interferences about physical libration and, in turn, the mass distribution in and internal geometry of the Moon. The picture we now have of the Moon is of a somewhat lumpy body much less spherical than the Earth. This may be because of lower gravity and higher rigidity. The center of mass of the Moon is 2.5 kilometers from the Moon's geometrical middle, toward the Nearside. The Moon's tidal bulge is 1,200 meters high. The Moon is only slightly oblate. The maria are 3 to 4 kilometers lower than the highlands. The Nearside is sprinkled with mascons. Finally, the Moon has discrete layers with different physical properties much as does its sister planet, the Earth.

One more word on the lunar laser ranging experiment: High-frequency variations in the lunar distance show that the Moon is vibrating as if it had been struck hard in the recent past. John Hartung has suggested that this is the legacy of an event witnessed by monks at Canterbury Cathedral in 1178 ("How High the Moon: A Decade of Laser Ranging," *Sky and Telescope,* October 1980, p. 274). During a summer evening of that year, a fiery glow was witnessed erupting from a horn of the crescent Moon. Hartung believes that this was the result of a rare event: the formation of a new large crater on the Moon. The reports of the Canterbury clergy are the only ones known to exist detailing an event such as this, and the "ringing" of the Moon that we can detect 800 years later lends them credence. It has even been suggested that the product of this event can be seen today on the Moon as the crater Giordano Bruno.

The reason that it is possible that the seismic activity generated by an impact that occurred in the twelfth century can still be detected today is that the Moon sustains vibrations to a much greater extent than does the Earth. This was one of the earliest discoveries made by the network of ALSEP seismometers established on the lunar surface to detect moonquakes. A tremor on the Moon will continue to register long after all traces of it would have been damped out on the Earth.

Moonquakes are much rarer than earthquakes. The Moon has only 1/100 of the seismic activity our planet experiences. Many of the moonquakes that are caused by shifting rock (not impacts) tend to occur well inside the Moon—much deeper than in the Earth. These moonquakes tend to occur when the Moon is nearest to the Earth, suggesting a tidal effect. They may be caused by shifts of material in the Moon while it becomes less pearshaped as it recedes from the Earth (and tidal forces become weaker).

A shallow type of moonquake is likely to happen at longitudes experiencing sunrise or sunset and may be the result of thermal stresses (F. Duennbier and G. H. Sutton, "Thermal Moonquakes," *Jour. Geophys. Res.*, **79,** 1974, 4351–4363).

Selenophysicists are interested in moonquakes because they are a way to probe the interior of the Moon. The way vibrations are transmitted through the satellite can be detected by using the multiple seismic detectors at the different Apollo landing sites. This information gives clues to internal structure.

The Earth's interior is divided into three parts: an outer crust, a plastic **mantle** beneath it, and a denser **core** at the center of the planet. Is the Moon differentiated in this way? The geologic and geophysical evidence says "yes," but not as much as the Earth.

In order to differentiate an initially homogeneous planet, one must have heat to melt it. If the Moon was once molten, where did the heat come from? One answer is that the Moon may have condensed hot out of a hot primordial nebula. If it formed out of cool material, though, the gravitational processes that brought the material together could also have produced enough heat to melt it. The kinetic energy of small bodies in space slamming into each other to form the Moon would have been transformed into heat. This **accretional heating** could have melted the material that eventually became the Moon.

There is another potential source of energy. The Moon may have been originally undifferentiated. Then, the radioactive elements in it, brought together in higher and higher densities by accretion, would begin to heat up. Some radioactive elements decay very quickly and would have released a great deal of energy in the early history of the Moon. One example is aluminum–26. Imbedded throughout the Moon, this energy source (along with longer-lived elements such as potassium–40) could have globally melted the Moon quite quickly. **Radiogenic heating** could have been responsible for the period of lunar volcanism later in the Moon's history, too. (It should also be kept in mind when we talk about liquifying the Moon's core.)

Both the accretional and radiogenic hypotheses conflict with some of the models for the origin of the Moon that we discuss in the next section. We do have evidence for at least some heating taking place in the Moon right now, though. Heat flow probes pounded into the lunar surface by the Apollo astronauts yielded positive results. The heat flow experiment, in fact, is cited as evidence of the enhancement of (radioactive) refractory elements in the Moon.

It is much easier to say what happened near the surface of the Moon than in the deep interior. If the Moon was not melted all the way through, it was at least covered at some time by a magma ocean, which may have been produced by accretional energy, radiogenic heating, or both. Such an ocean would lead to the fractionation of lunar materials. The lunar highlands were possibly an early crust floating on the magma ocean.

Many selenologists prefer the idea of a magma ocean to a totally liquid Moon. Some say the magma ocean was 200 kilometers deep. Others say that there was never really a global ocean but, rather, a series of magma lakes scattered about the primordial surface.

Currently, it is believed that the Moon of today does consist of a crust, a mantle, and possibly a core like that of the Earth. The lunar crust is quite thick—more than 70 kilometers deep on the average. It is probably even thicker on the Farside. Its thickness could have been the reason why magma from the mantle was not able to force itself upward and flood the Farside mare basins.

The lunar core is still an object of controversy. Hypothesizing a melted Moon and a lunar core is helpful in explaining the elemental abundances found on the Moon. Remember that the Moon seems to lack siderophilic elements. The Moon appears to have less siderophile material than either the Earth's mantle or the meteorites it is often compared to. Did they all sink into the lunar core so that we can't find them? The presence of a *small* lunar core (relative to the Earth's) may simply mean that differentiation was imperfect. A lunar core could be a receptacle for some missing volatile elements, too. Then, of course, there is the missing iron on the Moon. All these are constituents of the Earth's core. Yes, a lunar core would be quite convenient. Is there really evidence for such a core, or is this all wishful thinking?

A lunar core up to 500 kilometers in radius is consistent with the gravitational field of the Moon. Moonquake travel times (as noted by multiple ALSEP seismometers) hint at a core close to 300 kilometers in radius.

Another method for core detection relies on the magnetic properties that a predominantly iron core would have. It worked this way: A spaceprobe measured the solar and lunar magnetic fields near the Moon. At the same time an ALSEP instrument measured how much stronger the field was on the lunar surface (closer to the core). Scientists then calculated how much of the Moon would have to be core in order to obtain the difference between these two measurements. This method yielded a core size of 400 kilometers or less, which compares favorably with the core size obtained similarly by the Apollo XV and XVI subsatellite magnetometers.

If there is a lunar core, say 400 kilometers in radius and made up of iron and nickel, it will comprise only 2% of the Moon's mass (Figure 8–11). Thus, it cannot totally explain the scarcity of iron in the Moon.

The exact properties of the lunar core (just like its radius) are not known. For instance, it could be liquid or solid. On the Earth, the liquid core acts as a dynamo to produce a substantial **magnetic field.** The Moon, though, has a very weak magnetic field. Does a small core produce a

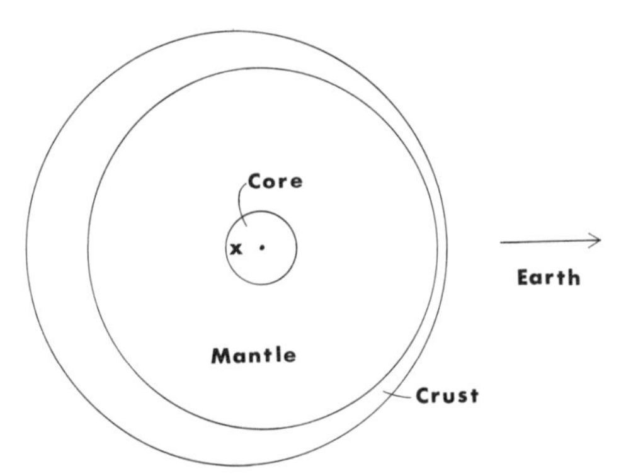

8–11 A cross section of the Moon. This stylized view represents the three layers in the Moon. The asymmetry has been emphasized. Notice how the crust is thicker on the Farside and that the core is displaced from the Moon's geometrical center (marked X).
(Andrea K. Dobson-Hockey)

small magnetic field? The answer is: Probably not, but it isn't impossible. For one thing, we don't know if the core is still hot enough to be liquid. We are left then with the question of where the weak magnetic field comes from.

The Apollo subsatellites detected magnetic anomalies on the lunar surface. It is possible that the shock of impact during crater formation may cause certain materials to become magnetic. Curiously, however, many of these magnetic anomalies do not correspond to heavily cratered areas. Perhaps large impacts have transported magnetic material as ejecta.

The Moon's magnetic state may have been more interesting in the past. Many minerals—such as troilite (iron sulphide)—retain a latent magnetism if they cool in the presence of a magnetic field. Lunar rock samples indicate that the Moon may once have had a substantial magnetic field, though once again, impact-induced magnetism could be fooling us. Still, if the **paleomagnetic record** reads true, there may have been 3.9 to 3.6 billion years ago a lunar magnetic field nearly equivalent to that of the Earth's today. (S. M. Cisowski, *et al.*, "A Review of Lunar Paleointensity Data and Implications for the Origin of Lunar Magnetism," *Proceedings of the 13th Lunar and Planetary Science Conference* in *Jour. Geophys. Res.*, **88,** 1983, A691–A704.) This implies that the lunar core would have been liquid at that time.

Liquid cores, elemental abundances, magma oceans, and heat sources for global melting all have implications when it comes to addressing one of the most fundamental questions in lunar science: Where did the Moon come from?

THE ORIGIN OF THE MOON

On October 3, 1984, lunar scientists gathered in Kona, Hawaii, to discuss the origin of the Moon. Only then, 12 years after Apollo, was enough known about the Moon and, equally important, the Earth to devote a three-day meeting to the subject. Several competing theories were examined. Scenarios thought up in the past century were put head-to-head against new hypotheses developed in recent years. All had their proponents. Arguments for and against each were complex, lucid, and sometimes vehement.

In 1755, Immanuel Kant (1724–1804) elucidated a theory of the formation of the solar system. Elaborated by Laplace, this theory sees the primitive solar system as a nebula of gas that by virtue of internal forces collapsed into a disk and then condensed into the Sun and planets about 5 billion years ago. What triggered the condensation is unclear. It may have been the gravitational interaction of a passing star or a nearby **supernova** explosion.

This is all well and good for the planets, but what mechanism caused such a large body as the Moon to be orbiting the Earth? Classically, there are three theories of the origin of the Moon: **fission** (the Earth is the parent body of the Moon), **capture** (the Moon was once an errant body in the solar system), and **coaccretion** (the Moon formed independently near the Earth). In addition, there are many newer variants and hybrids of these older theories.

One would think that it would be a simple matter to run back the orbital history of the Moon and see where that body was 4.5 billion years ago. If this playback in reverse time found the Moon elsewhere in the solar system, the capture theory would win. If it stayed pretty much in the neighborhood where it is, coaccretion would be the answer. If it crashed into the Earth, the fission adherents would have their day. Unfortunately, it's not that easy.

The study of the Moon's orbit wound down during the twentieth century. However, when a practical need was felt with the coming of lunar exploration by space probe, it was necessary to start studying it intensively again. New techniques, such as laser ranging and extrapolation from the behavior of artificial statellites, were brought to bear on the problem. In addition, older methods such as stellar occultations and ancient eclipse records were re-examined. (More exotic is the study of fossilized sea creatures, which record the length of the day and month in the layers of their shells.)

It has been found that although we can say with great precision what the Moon is doing right now (for instance, moving away from us), its pattern of movement is changing. Furthermore, this change occurred more rapidly in the past. More than 70% of the Moon's orbital evolution took place in the first half of its life. Thus, most of the interesting things that have happened to the Moon did so long ago when we know least about it.

We do not know where the Moon started. We can go back only so far and state the problem this way: How did the Moon approximately 4.5 billion years ago come to be orbiting the Earth from west to east at a distance equal to 10 Earth radii and at an orbital inclination of 10 degrees?

The Moon is unique in the solar system. The Earth–Moon system contains more angular momentum per unit of mass than any other such system. Unlike other planets and satellites, the Moon has a small core. We can add to its list of peculiarities the Moon's unique chemical composition—not quite Earth-like, not quite like other rocky bodies in the solar system—and the fact that it has survived whatever history it has experienced intact with a more or less constant volume and without any signs of great stress.

Darwin believed that the Moon was unusual and did not originate in the same way as the other planets. Because of his work with tidal interaction, he was able to extrapolate the evolution of the Moon backward to a time when the Moon was much closer to the Earth and when the month and day were nearly the same length. He concluded that the Moon and the Earth were originally one body. In 1878 he said that as the newly formed hot Earth cooled, it contracted. To conserve angular momentum it spun up. When the still-liquid Earth's rotation period was between three and five hours, the solar tide caused an oscillating bulge to form at the equator. This situation was unstable, and finally the bulge tore off to become a separate body, the Moon. This body cooled and became a satellite of the Earth. It has been moving away from its parent body ever since.

Darwin's theory was later embellished by the suggestion that the Pacific Ocean is the scar left over from this violent fission (it's about the right size), though it is unlikely that such a scar would remain from the separation of two liquid bodies. Subsequently, the theory of continental drift has made this point moot by showing that the Pacific Ocean has existed only relatively recently in geologic time.

Classical fission was seemingly put to rest in 1930 by Harold Jeffreys, who demonstrated that the viscosity of a molten Earth would damp out the oscillations necessary to produce the proto-Moon bulge. (Stephen G. Brush, "Nickel for Your Thoughts: Urey and the Origin of the Moon," *Science,* **217,** September 3, 1982, 891.) The theory has recently been revived, though, by Alan Binder and others who point to binary star formation as an analogy for lunar fission. Some stars are indeed thought to spawn other stars, but this process is different from that proposed for the Moon.

A handicap of the fission theory is the tremendous amount of energy required. Also, classical fission would require a totally molten Moon, a requirement not all selenologists are willing to accept. Even if fission occurred, it is still necessary to explain why the Moon is not orbiting its primary in the plane of the Earth's rotation.

The Moon's low density is a point in favor of the fission theory. If the Earth was differentiated before fission, the Moon would have been torn out of that body's mantle, not its core. Thus, the Moon would be deficient in siderophiles and lack the iron to form a substantial core. Alternately, fission might have occurred before the Earth had finished forming. The volatiles would have been among the last elements to condense out of the solar nebula, and the Earth's mantle may not have received its share of them before giving birth to the Moon. (H. Wänke, "Constitution of Terrestial Planets," *Phil. Trans. Roy. Soc.,* **A303,** 1981, 287.)

Likewise, because the Moon was necessarily formed hot if it was formed by fission, the easily vaporized volatiles could have been lost in the process.

In order to overcome the physical difficulties with

fission while keeping its advantages in producing a Moon similar to the Earth's mantle, it has been suggested that the Earth spun off the Moon as a stream of material (much like water off a rotary garden sprinkler). In the 1970s, A. Edward Ringwood and A. G. W. Cameron independently theorized that the Moon condensed out of an extended protoatmosphere of the Earth ("The Origin of the Moon," *Science,* **216,** May 7, 1982, 606). Once material was carried to a distance from the Earth, it might have formed a disk or ring around the planet. Some material would have been lost to space, of course; some would have been trapped within the Roche limit. Still, there could have been enough left to condense and form the Moon.

Formation of the Moon from a ring of material is a variant of the binary, or coaccretion, theory. Coaccretion in its purest form is a truly double planet scenario: the Earth and the Moon formed on equal footing. There just happened to be two centers of condensation in the solar nebula near each other, and when one grew more slowly or for a shorter period than the other, it became the other's satellite. It could have been the other way around: the Earth could have become a satellite of the Moon!

This idea is old. Laplace and Gilbert addressed it, and Edouard Roche (1820–1893) dealt with it quantitatively. More recently, Kuiper proposed a coaccretion model.

The theory has the ratio of the sizes of the Earth and the Moon going for it, but not much else. After coaccretion, the Earth and the Moon would seemingly have been identical except for size. All subsequent evolutionary differences would have to be accounted for by the disparity in mass alone.

One way to counter objections to the theory is to say that the Moon started forming *after* most of the iron in the solar nebula had condensed out. All the iron in the Earth–Moon viccinity would have been used up by the Earth. By the time the volatiles condensed out, the Earth would already have been more massive than the Moon, and therefore it could have competed more successfully for gleaning the remaining elements (W. A. Cassidy, "The 'Problem' of Iron Partition between Earth and Moon during Simultaneous Formation as a Double Planet," abstr., *Papers Presented to the Conference on the Origin of the Moon,* 1984, p. 45). This gives us some of the chemical differences between the Earth and Moon that are needed to support the theory, but it is somewhat *ad hoc.*

Completely separate from the fission and coaccretion traditions, another theory of lunar origin developed in the nineteenth century. Capture of the Moon by the Earth was first advocated by Thomas See (1886–1962). Its greatest modern proponent was Urey. The classical capture theory says that the Moon formed as a planet in its own right somewhere in the solar system and then passed near the Earth, where it was gravitationally captured. Capture played third fiddle to fission and coaccretion for a long time, but immediately after Apollo it seemed that at least some version of capture might be likely when the moonrocks were found to be significantly different from the terrestrial mantle. This would be hard to explain if they had once been one and the same or even if they had been formed close to each other. If the Moon formed at a different place in the solar system, it would not have to have the same makeup as the Earth.

There is evidence for gravitational capture elsewhere in the solar system. Examples are several of the satellites of Jupiter that are thought to be asteroids the giant planet captured. Triton may have been an independent body before it met up with Neptune.

It is certainly feasible that the Moon formed someplace else and then traveled by the Earth. The problem is getting it to stay there. Remember, the Moon is no tiny asteroid, and the Earth is not a giant planet. It is one thing to catch a baseball; it is quite another to catch a medicine ball.

How could the Earth have captured the Moon? Tidal interaction is possible—just barely. Through a complicated sequence of multiple passes and strange orbits that make computers heat up just to calculate them, it is possible for the Earth to wrestle with the Moon and make the pin. This necessitates that the Moon come in at just the right speed and direction at just the right time. The probability of this happening is quite low.

To make the possible trajectories of the Moon before capture more plentiful and therefore the event itself more likely to occur, we must invoke methods other than tides for dissipating the Moon's energy. Suppose that the Earth had had the extended primitive atmosphere that I talked about earlier. If the incoming Moon had had to wade through it, drag would have slowed the would-be satellite. In one scenario, one would have to be careful that the Moon did not simply rip away the atmosphere and keep going, but under the right circumstances, it could work. Of course, this atmosphere could not keep on slowing down the Moon. If it still existed after it had caused the Moon to fall into orbit, it would continue to impede the Moon. Eventually, the Moon would spiral into the Earth and be no more. The envisioned extended atmosphere then would have to dissipate quickly once the Moon became a satellite.

Alternately, the approaching Moon could be slowed by striking small bodies already in Earth orbit. We will tackle this idea in another context later.

For the foregoing capture encounters to work, the Moon must be traveling initially in an orbit somewhat similar to that of the Earth's. If it comes streaking in toward the Earth from the far reaches of the solar system, it will be traveling too fast to stop in any circumstance. This implies that a captured Moon would still need to be formed at about the same distance from the Sun as the Earth was. How then do we explain the chemical inconsistencies? By trying to solve the dynamic problems of capture, we are in danger of throwing away the geochemical arguments in its favor.

So far we have implied that the Moon is captured *intact*. That is, tidal interaction does not tear it apart, which is certainly a danger if the Moon stays within the Earth's Roche limit for very long. Maybe this is what we want, though. If the capture is *disruptive,* perhaps the fragments will take to orbiting the Earth more easily than will the whole body, and the Moon will reassemble in orbit by accretion at its leisure. This won't happen on one encounter; a body passing rapidly through the Roche limit will just sail by unharmed. Only if it lingers there will it be torn asunder. Still, a body that has multiple encounters with the Earth might be broken up, and disruptive capture at least gives us a little more latitude in the possible ways in which the Moon might show up on the scene.

To the three theories of lunar formation let us now add a fourth. Each of the classical theories in some way invokes a ring of material in orbit around the Earth accreting to form the Moon (This idea has been favored by Soviet selenologists for some time.). Material captured into an **accretion disk** could explain the geochemical nature of the Moon. Suppose a body was tidally disrupted by an encounter with the Earth. Being a typical rocky body, it is differentiated. The resulting pieces that were the crust and mantle will be less dense than those that were the core. The core may even stay intact. It will be less likely to be gravitationally captured by the Earth. A lesser amount of energy is necessary to capture the crust and mantle pieces. Thus, we have a mechanism for sifting out just the material that we want for our Moon: The silicates form a prelunar disk around the Earth while the core keeps on going and is lost. A body about the size of Mars would do the trick so that a mass about that of the Moon's is left behind.

John Wood and Henri Mitler have proposed that instead of one body, the Earth captured a whole string of **planetesimals**—small differentiated bodies. The same thing would happen. The bodies would be tidally "shelled" (Alan Rubin, "Whence Came the Moon", *Sky and Telescope,* **68,** No. 5, November 1984, 389).

A number of lunar scientists believe that existing material in Earth orbit would act as a filter on the planetesimals. Metal-deficient parts of planetesimals would be preferentially gathered into a swarm and further sorted by repeated collisions with their neighbors. The cores would pass through the swarm. (Clark R. Chapman and Richard Greenberg, "A Circumterrestrial Compositional Filter," abstract., *Papers Presented to the Conference on the Origin of the Moon,* 1984, p. 56). The planetesimals that would pass through the filter would be heliocentric orbiting bodies. Those orbiting outside the Earth's orbit would be traveling more slowly than the Earth; the Earth would overtake them. Those orbiting inside the Earth's orbital radius would be traveling faster and would overtake the Earth. The net amount of angular momentum that would be added to the system would be zero (Figure 8–12). Thus, we have yet to resolve the problem of why the Earth–Moon system has so much angular momentum. We need some way to add this unaccounted-for angular momentum to our equation.

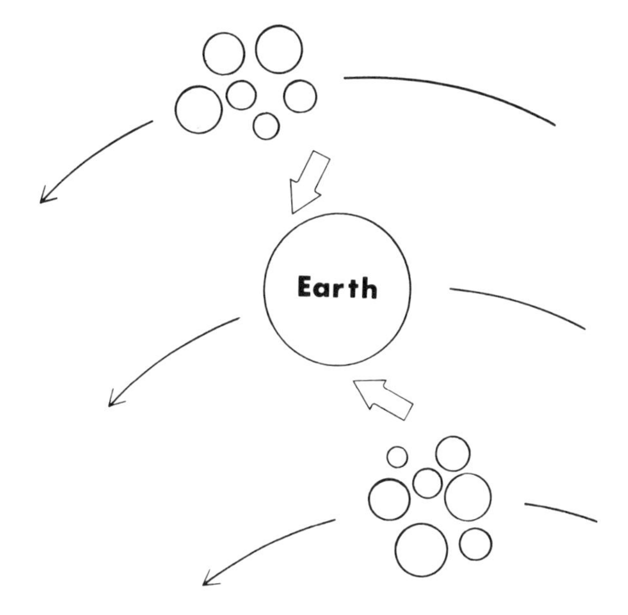

8–12 Planetesimals impacting the Earth. Planetesimals orbiting closer to the Sun than the Earth will overtake the Earth and impart a clockwise rotation to it upon impact. Planetesimals orbiting farther from the Sun will be overtaken by the Earth and will impart a counterclockwise rotation. On the average, these two effects should cancel each other.
(Andrea K. Dobson-Hockey)

The Tucson Lunar Origin Consortium (Michael Drake, Floyd Herbert, and John Jones of the University of Arizona's Lunar and Planetary Laboratory and Richard Greenberg, Clark Chapman, Donald Davis, William Hartmann, and Stuart Weidenschilling of the Planetary Science Institute) believes that the Moon was formed by the **impact** of one or more bodies into the Earth. The impact hypothesis was formulated by William Hartmann and Donald Davis and also—independently—by A.G.W. Cameron and William Ward. The new hypothesis borrows elements from each of the older theories of lunar origin. In the new hypothesis, a differentiated planetesimal or planetesimals strike the Earth obliquely, and the residue of the collision is deposited in orbit. If the planetesimal simply bounces off the Earth, this idea resembles the capture theory; if the impact tears loose a great deal of the Earth's mantle which in turn becomes the Moon, impact becomes another fission mechanism. However, if it is suggested that the Moon is made up of both exogenic material and debris thrown up from the Earth's mantle, the impact hypothesis gives selenologists a way of explaining the lunar composition.

Take, for instance, the core. If a planetesimal core shed its mantle on impact, the core could have skipped back on into space. Alternately, it could have been captured by the Earth and caused to crash. The core might then have sunk into the molten proto-Earth to join with the Earth's core. (It might have done this at initial impact, too (see A. G. W. Cameron, "Formation of the Pre-Lunar Accretion Disk," abstract., *Papers Presented to the Conference on the Origin of the Moon,* 1984, p. 58). Either way, the core is lost, and the resulting lunar constituents are iron-deficient.

Imagine now that at the time of the formation of the solar system, an extra body about the size of Mars exists. (Computer models say that this is possible.) It travels through the solar system until it comes near the Earth. The set of possible trajectories that it might take is broader now because we no longer require that it slip into a permanent

Earth orbit. It might only be captured by the Earth and eventually drawn toward it. Finally, the body hits the Earth. The recoil sends material flying upward. By themselves these chunks would eventually fall back to the Earth. Any number of smaller planetesimals may have struck the Earth in its distant past, and we will never know it. However, a large impact involves so much energy that a fine mixture of fragments and mantle is produced. Much of it will fall back, too; some will reach escape velocity. However, enough will be inserted into orbit to accrete and begin to form the Moon. In fact, it may do this quite rapidly.

Why doesn't all this material fall back down? Perhaps there is already a ring of material circling the Earth (from the capture of planetesimals?) with which the impact ejecta interacts. Another explanation is that the impact energy vaporizes the material and produces a hot silicate atmosphere around the Earth (around 2,000 K). Pressure effects within this gas might give it the extra energy necessary to achieve orbit.

What is the likelihood of a collision between the Earth and a large planet-sized body? No one knows for sure. There is evidence of large-scale collisions elsewhere in the solar system, though. Phobos, a satellite of Mars, shows the scars of an impact that may nearly have destroyed it (Figure 8–13). Satellites in the Jovian and Saturnian systems have many huge crater-like features on them. Mimas (orbiting Saturn (Figure 8–14) may even have been totally fragmented by a collision and then reassembled. We don't see really large bodies flitting about the solar system today (though there are a lot of asteroid-size bodies out there). If they were once plentiful, there must have been impacts. Have they all been used up?

The Moon's unique composition is probably the most positive evidence for the impact hypothesis. The impacting body could be traveling in a very eccentric orbit around the Sun. Thus, its composition could have been determined by a variety of conditions that existed in the

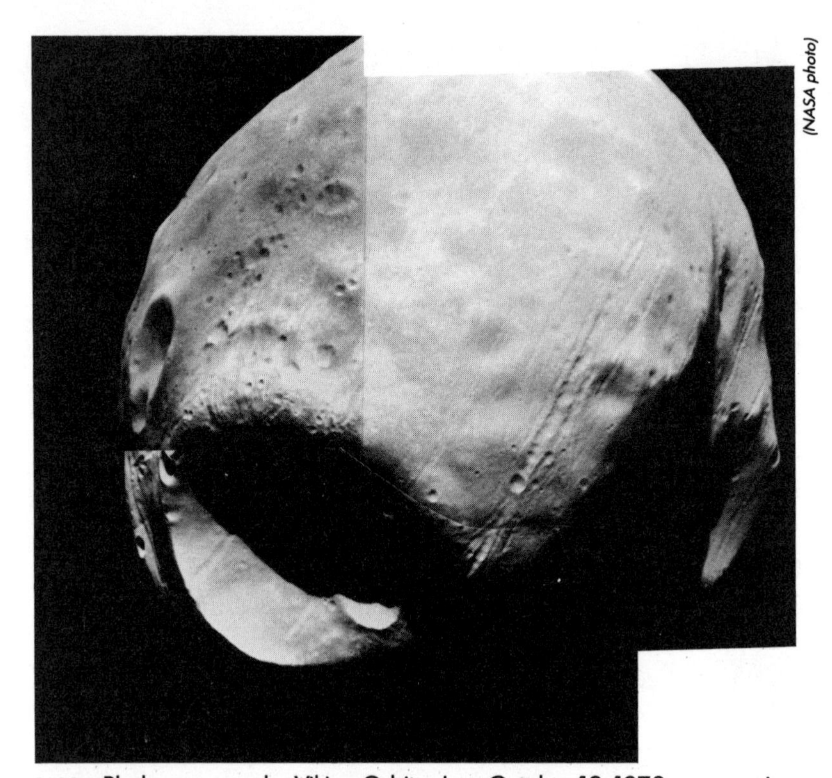

8–13 Phobos as seen by Viking Orbiter I on October 19, 1978, appears in this composite photograph. The satellite is 612 kilometers away. The largest crater, Stickney, is 10 kilometers in diameter. The linear grooves coming from and passing through the crater may be fractures caused by the tremendous impact event that formed it.

early solar nebula. During impact, volatiles would have been blown away from the vaporized mix. We have already said that the resulting material would be depleted of heavy elements. Thus, by mixing proto-Earth mantle with material from a body formed elsewhere and perhaps adding in wandering planetesimals captured by the impact-induced accretion disk, it is possible to get nearly any chemical composition desired!

Ironically, returning once more to the angular momentum argument with which we introduced the impact hypothesis, the "kick" of a large impact may not be enough to explain all the high angular momentum in the Earth–Moon system. Planetesimals might add more angular momentum now and then (the cross section for their capture would be

8–14 Mimas, photographed by Voyager I on November 12, 1980, from a range of 425,000 kilometers. The huge impact crater is more than 100 kilometers in diameter.

increased by an impact-induced accretion disk encircling the Earth), but we have already decided that this would tend to cancel out. Too, the impact hypothesis, like the others, still does not tell us why the Moon has the particular orbital inclination it has today.

The impact hypothesis is popular now among some lunar scientists because it is a synthesis and partly, perhaps, because it is new. It requires invoking a stochastic event governed by the rules of chance. Still, much of modern physics is based on the laws of probability. One thing is certain: The impact event necessary to produce the Moon was not a likely happening in the sense that the odds would not have favored it. Thus, a satellite produced in this way should be unique in the solar system. Another system just like the Earth–Moon system would stretch credibility. It is interesting to think that the enigmatic Pluto system in the farthest reaches of the solar system may have something to tell us about the possible way the Moon was formed.

No consensus was reached at the Conference on the Origin of the Moon. None was intended. Instead, scientists called for more data on the problem. Much time was spent discussing the experiments that will need to be performed and the samples that will have to be taken in order to shed more light on the lunar origin problem when (not "if") we return to the Moon.

9
TOWARD THE YEAR 2007

BEYOND APOLLO

After the initial Apollo landings established the where-with-all to go to the Moon, lunar exploration in future missions was to be extensive. A flight would land on the rim of Tycho near Surveyor VII. Another would visit the Marius Hills, the site of possible volcanic domes. Apollo XVIII was supposed to touch down in Schröter's Valley near Aristarchus. Apollo XIX was to explore a crater chain, and finally Apollo XX was to land inside the crater Copernicus. It was hoped that additional flights, if authorized, would go to the lunar North Pole or even the Farside.

None of this was to be, however. By the late 1960s, the NASA budget was already being cut to support the war in Vietnam. After Apollo XI, the goal of being first on the Moon had been accomplished. The novelty wore off, and public interest in Apollo waned. People called for using the money spent on Moon missions to try to alleviate domestic problems.

The year after the first lunar landing, Apollos XVIII through XX were cancelled. The Apollo Applications Program, a plan to use Apollo hardware for a variety of tasks near the Earth and the Moon, was reduced to the Skylab missions in Earth orbit (Figure 9–1). (The last Apollo CM/

SM was used in 1975 for the largely political Apollo-Soyuz Test Project in a joint flight with the Soviet Union.)

Interest in the Moon deteriorated so rapidly in the 1970s that in 1978 the remaining Apollo ALSEPs on the Moon were shut down because of lack of money to monitor them. At the Lunar Receiving Laboratory funding cutbacks curtailed the opening of new sample boxes. A NASA report

9–1 The Skylab space station.

(NASA photo)

revealed that a small amount of the lunar samples might have actually been mislaid! Accounting procedures were blamed.

Meanwhile, what had happened to the Russians? In the 1960s it was believed by almost everyone that the Soviet Union was attempting to land a man on the Moon and to do so before the United States. This possibility spurred on the Apollo program as it counted down toward the end of the decade. It was puzzling that the Soviets had failed to try by the time of Apollo XI. They certainly appeared to be near possessing the necessary technology, having led the United States in space achievements for most of the first decade of space flight. Then, the USSR made it known that they had never really intended to beat the United States to the Moon in the first place! At first this sounded like sour grapes; Russia observers (like NASA director Thomas Paine) still expected them to be there within 18 months. However, the Soviets remained in Earth space during the 1970s, directing their attention to constructing an Earth-orbiting space station, and their explanation about the Moon gained credibility. It began to look as though the space race had never been real. It was even suggested that it was part of a Soviet trick to direct American money and effort away from defense-related projects. More recently, it has been proposed that the Soviet Moon program was genuine after all (Peter Pesavento, "Soviets to the Moon: the Untold Story," *Astronomy,* **12,** December 1984, p. 6). Certain quotations from cosmonauts at various times during the past 15 years bear this out, but were these comments just propaganda?

The best evidence that the USSR was indeed planning to send cosmonauts to the Moon centers around the Zond series of spacecraft. Zond was a large space probe and resembled part of the most recent model of Soviet manned spacecraft, the Soyuz. A Zond had capabilities that are unnecessary for an unmanned vehicle. It had a **launch escape system** designed to remove a spacecraft from atop a booster if it was in imminent danger of explosion on its launch pad. (This was a staple feature on all Apollo launches.) Also,

Zond's re-entry from the Moon was designed to keep deceleration forces below the maximum tolerable for a human being. The Zond spacecraft orbited the Moon and returned photographs to the Earth, but was it a prototype for future manned lunar missions?

Although the Russians demonstrated the capability to go to the Moon, it is unclear whether they could have landed there. There is no direct evidence of the existence of a Soviet spacecraft equivalent to the U.S. lunar module. However, Soviet cosmonauts conducted flight practice aboard helicopters. A helicopter would be a likely training device for a lunar lander. The USSR may have tested such a lander in space during 1970 and 1971 as part of the mysterious Cosmos series of orbiting satellites.

Writer Peter Pesavento goes so far as to suggest that the Zond VII mission of August 1969 was actually intended to be a manned Moon-orbiting mission. It may have been scheduled to fly in tandem with a lunar sample retrieval flight. With the two flights under its belt, the Soviet Union could have claimed that it had sent men to the Moon *and* brought back lunar samples, thereby scooping the United States had the first Apollo landing been delayed until Apollo XII (a distinct possibility at the time). Pesavento believes that this plan was scrubbed after the Russian automated space probe blew up on the pad.

How close the Russians were to getting on the Moon may remain cloudy for some time. One thing is certain: If they did have a manned Moon program in operation, they dismantled it quickly after the successes of Apollo. The USSR was not interested in being No. 2 on the Moon. (Recently, the Russians have publicly speculated about a future manned Soviet landing on Mars.)

Although the history of the Soviet Union's manned lunar program is uncertain, its unmanned lunar exploration is a fact. It continued longer than that of the United States. A Soviet spacecraft did visit the Moon in July 1969. It flew concurrently with Apollo XI, and speculation was rampant about its mission. Was it somehow intended to take away some of the attention being lavished on the American flight? It traveled slowly: Was it carrying an extra heavy payload, or was it economizing on fuel for a possible landing? The Russians, as usual, said nothing. Luna XV orbited the Moon and then changed that orbit after two days. It finally crashed. Luna XV may simply have been a test of maneuvering spacecraft in lunar orbit. It may have been sent to study the mascons.

On the other hand, Luna XV may have been a failed automated mission for sample return designed to upstage Apollo XI. This argument gained credence when future Luna series spacecraft did attempt sample returns. Luna XV was the same weight as Luna XVI, which succeeded in returning samples. Luna XVI landed in Mare Fecunditatis in September 1970. It and its later counterparts collected smaller and less discriminate samples from the Moon than did Apollo, but they managed to retrieve rock from areas unexplored by the American missions, which aided our global understanding of the Moon. ("A Decade of Moon Rocks," *Sky and Telescope,* July 1979, p. 11).

Luna XVII deposited a unique automated device on the Moon in November 1970. The Lunokhod vehicle was able to travel over the lunar surface on an extensive journey of exploration. In August 1976, Luna XXIV made a 2-meter bore hole in the Moon.

The year 1979 marked the tenth anniversary of the first manned landing on the Moon. There was little celebration. It was also the tenth anniversary of the Lunar and Planetary Science Conference (LPSC). (Originally, the LPSC was simply The Lunar Science Conference, but the name was changed in 1978 to reflect the influence of data from other planetary space exploration programs.) The Lunar Science Conference had been meeting yearly ever since the first one that had been held only six months after Apollo XI. At these gatherings, lunar scientists from all over the world kept the flame of lunar exploration burning through the indifferent 1970s. Each year more information gleaned during the previous 12 months from the moonrocks and other

Apollo data was revealed. Sometimes this was made possible by new technology; other times it was the result of long-term research projects only then coming to fruition.

While U.S. manned spaceflights became devoted to the reusable Earth-orbiting Space Shuttle, unmanned exploration continued to push out into the solar system. Pioneer, Mariner, Viking, and Voyager spacecraft sent back pictures and other data from Mercury, Venus, Mars, Jupiter, Saturn, and their satellites. Instead of two worlds to compare, there were now close-up views of more than a dozen for planetary scientists to examine. Information gained from the *in situ* exploration of the Moon was found to be beneficial to analysis of these other planets as well.

No other body in the solar system more resembles the Moon than does the planet Mercury (Figure 9–2). Photographs of its cratered surface are almost indistinguishable from those of the Moon. There are differences, however. Mercury's greater gravity has limited the range of ejecta and impeded the formation of some large features. Mercury exhibits a global tectonic pattern that shows that it underwent tremendous adjustments in its crust at one time. As its rotation rate slowed, the planet probably became more spherical. Mercury also has a magnetic field.

The planet Mars is in some ways intermediate in appearance between the Moon and the Earth (Figure 9–3). Its thin atmosphere reveals impact craters that are eroding away only very slowly. There is little water on Mars. However, Mars also exhibits sand dunes formed by blowing wind, and there are signs that it had flowing water long ago. Venus, on the other hand, with its thick cloud-covered atmosphere and possible shifting land masses, is much more Earth-like than Moon-like (Figure 9–4).

In the outer solar system, planets like Jupiter and Saturn have satellites larger than our Moon. These satellites are unlike the rocky Earth and Moon or the gas giants that they orbit. They are bodies primarily made up of ice, yet in some ways their surfaces appear familiar. Craters can be seen on a satellite like Ganymede (one of the Galilean satellites),

(NASA photo)

9–2 The planet Mercury from a mosaic of Mariner X images, taken on March 29, 1974, when the spacecraft was 200,000 kilometers away. The largest craters are approximately 200 kilometers in diameter.

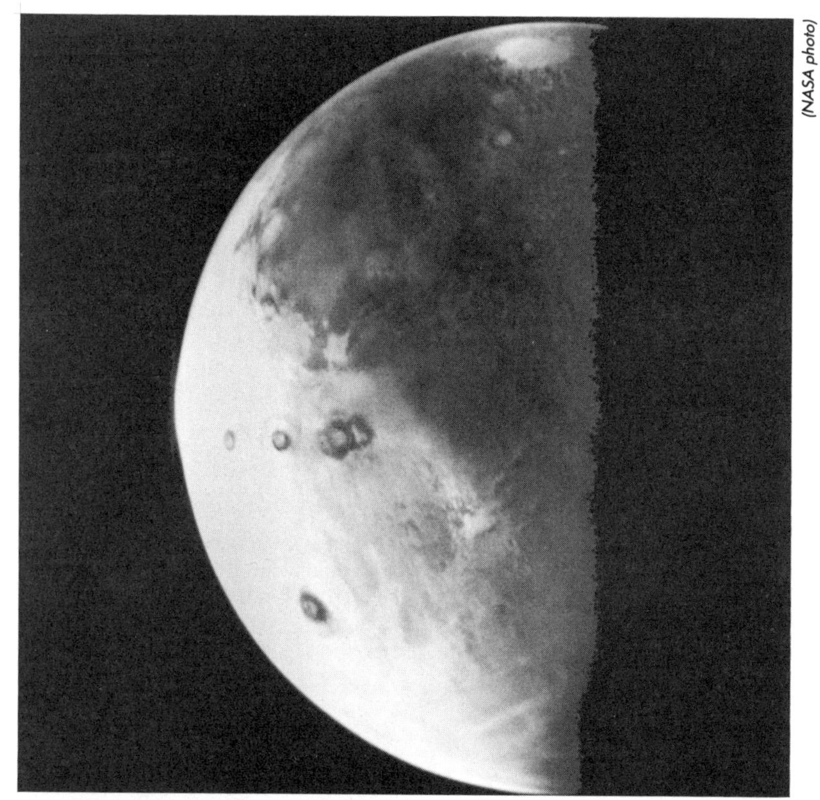

(NASA photo)

9–3 Mars as viewed from 560,000 kilometers away. A row of immense volcanoes can be seen across the middle of the half-disk. Near the bottom are several impact craters. This photograph was taken by Viking Orbiter I on June 17, 1976.

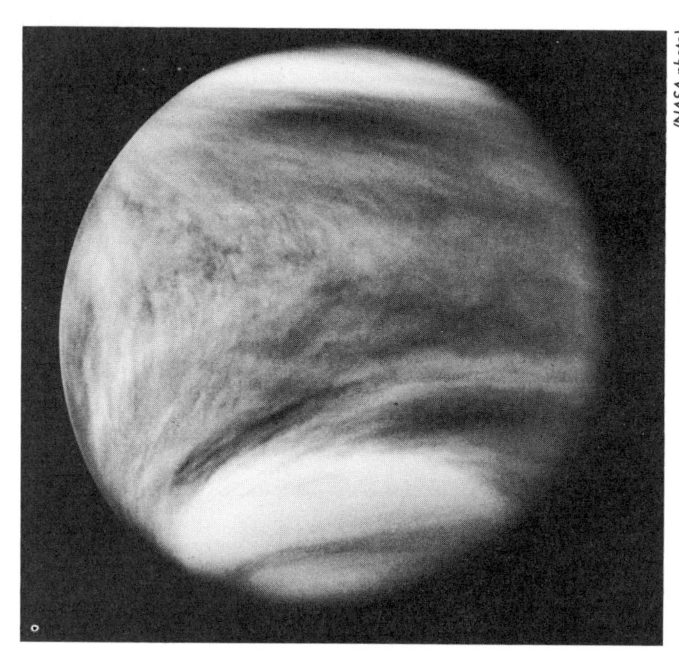

(NASA photo)

9–4 The cloud-swept disk of Venus seen by the Pioneer Venus Orbiter in 1979.

for instance. Ganymede experienced a period of heavy bombardment just as the Moon did. Some of its craters have rays. There are no central peaks, though. On Ganymede, the larger the crater, the shallower. Low impact "basins" show no relief at all because the icy crust in which they are formed flows. Thus, gravity pulls down crater rims after a period. Craters are a transient phenomenon on Ganymede (Figure 9–5).

There are rille-like cracks on Ganymede, too. These show up as linear expanses of high-albedo sub-surface ice that have been exposed by tectonic processes.

9–5 The surface of Ganymede photographed on March 5, 1979, by Voyager I from a distance of 246,000 kilometers. The smallest features that can be resolved are about 2.5 kilometers across. Note the extensive ray systems of the impact craters. The lighter circular areas in the upper right portion of the photograph may be ghost craters.

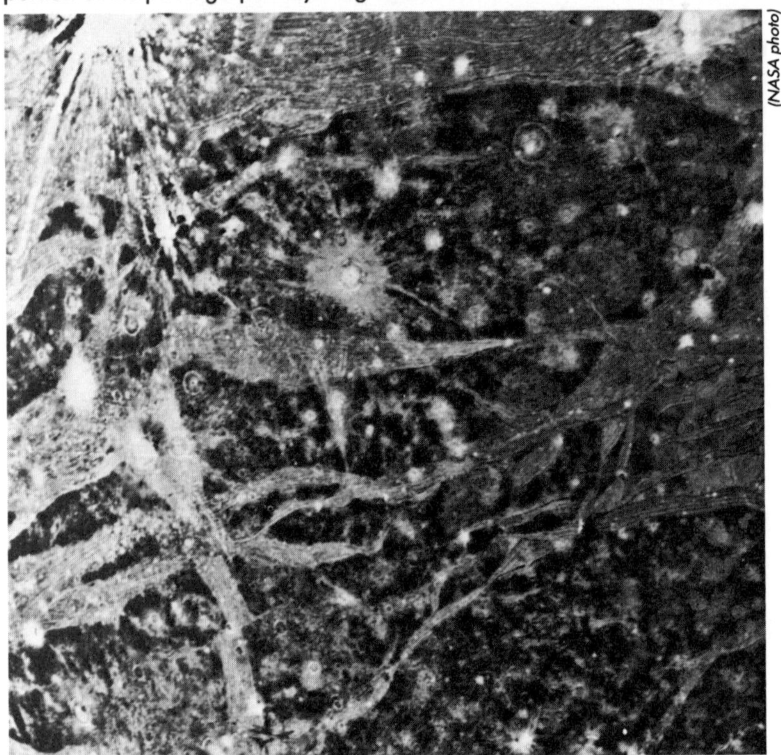

(NASA photo)

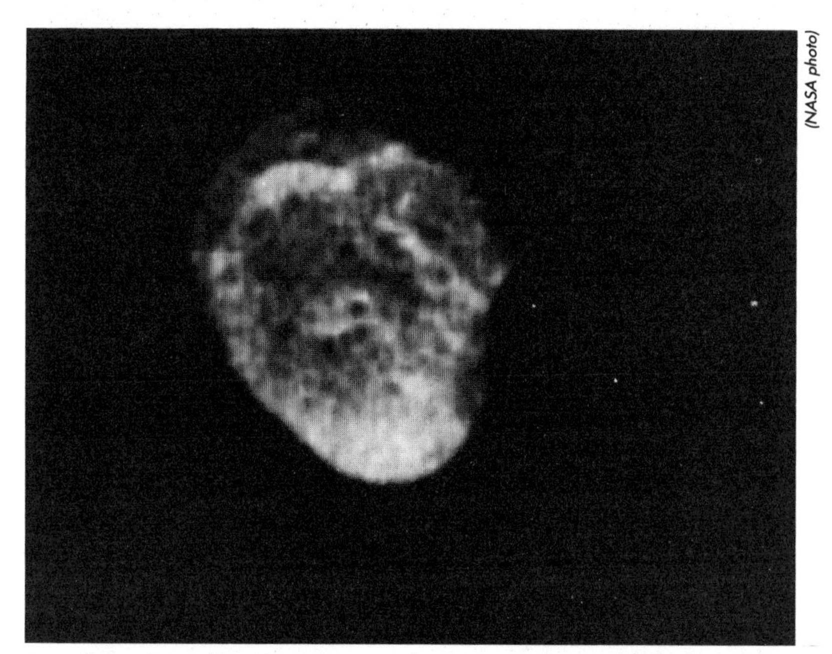
(NASA photo)

9–6 Saturn's satellite, Hyperion, may be a captured asteroid. Its distinctly non-spherical shape can be seen in this Voyager I picture taken on August 24, 1981, from a distance of about 500,000 kilometers.

It is not necessary to send out space probes to study another class of bodies. Meteorites are thought to be similar to the asteroids (Figure 9–6). Unlike on the Moon, meteoroids that fall to the Earth as meteors are slowed by the atmosphere. Some survive intact and can be collected as meteorites.

Scientists go to Antarctica to find quantities of meteorites. In the middle of that land mass surrounding the South Pole there is little erosion. In the mid-1970s it was discovered that meteorites found there were often considerably older than those found elsewhere and that they have accumulated. Every year now there is an expedition to that cold continent to retrieve them. ("Message from the Moon," *Scientific American,* 250, February 1984, 76.)

Scientists were shocked upon returning from the 1981 expedition when they found that one meteorite looked very familiar. Its composition did not match that of the asteroids. Rather, it resembled the specimens found in the sample cases at the LRL! The meteorite, called "Allen Hills 81005," even *looked* like a moonrock. It appeared to be mostly a breccia of fine-grained minerals. ("Message from the Moon," *Scientific American,* 250, February 1984, 76). These minerals are rare in meteorites but common in the samples of Apollo. Also, Allen Hills had been exposed to the solar wind for some time, which could not have happened on the Earth. (Richard A. Kerr, "A Lunar Meteorite and Maybe Some from Mars," *Science,* 220, April 15, 1983, 288).

The 31 gram rock, 3 centimeters in diameter, was geochemically identical to one found on the Moon, but how could a moonrock get to Antarctica? This discovery might have been invoked as evidence of ancient astronauts five years earlier when such mythology was popular. Today, it is thought that Allen Hills was a surface lunar rock that was accelerated to escape velocity by an impact. It then fell to the Earth as a meteor. Models have been put forth to explain why an impact with that much energy did not melt or destroy the errant fragment. Allen Hills is remarkably intact; the breccia is not deformed. (Richard A. Kerr, "A Lunar Meteorite and Maybe Some from Mars," *Science,* 220, April 15, 1983, 288).

Is it possible to find the spot where Allen Hills was chipped off the Moon? Its composition is that of a highlands rock, but its basalt content indicates that it may also have been near a mare. Allen Hills has very little KREEP. KREEP is primarily a Nearside phenomenon, and a lack of KREEP in quantity suggests that the meteorite originated on or near the Farside. It could conceivably even be from the event that produced the new crater Giordano Bruno. (J. Eberhart, "Moonrocks Yes, Marsrocks Maybe," *Science News,* 123, March 26, 1983, 196.) If this is true, Allen Hills gives us yet another data point on the lunar surface: Allen Hills contains certain minerals that are not found in the Apollo samples. As a sample it is therefore invaluable.

The arguments for the genesis of Allen Hills are being followed closely by planetary scientists who believe that another Antarctic meteorite may be from Mars. Meanwhile, the sample collecting goes on on the ice fields of Antarctica—as unusual a place as there is to practice selenology.

During the period of deep-space exploration in the 1970s and 1980s, it was suggested that some of the new technology developed for the purpose be directed back at that old friend of planetary science, the Moon. In the 1970s a proposal was made for the Polar Orbiting Lunar Observatory (POLO). POLO was to be advanced, unmanned, lunar observation platform. A technological generation beyond Lunar Orbiter, it would not simply take pictures of the Moon, it would geochemically map it. The emphasis of POLO was to be determination of the resources of the Moon on a global scale—seeing what it has to offer. In this sense, POLO was not just an upgraded space probe for lunar exploration; it was a scout for future economic development of the Moon.

POLO was to use some instruments similar to those carried on the Apollo subsatellites, but its polar orbit would take it over all the Moon, not just the equatorial regions. Its low orbital altitude and three-axis stabilization would afford a perfect platform for assessment work.

POLO would have been expensive, though—not as expensive as virtually any deep-space probe, but expensive enough—and most of NASA's meager budget was going toward the Space Shuttle. Planetary research funding was at an all-time low. POLO was originally a joint project between NASA and the cooperative European Space Agency (ESA). However, NASA was forced to pull out, and ESA went on alone. ("ESA Studying All-European Lunar Orbiting Laboratory," *Aviation Week and Space Technology,* 116, February 15, 1982, 124). Eventually, POLO was shelved.

Almost immediately, a plan for a more modest POLO-like spacecraft called Lunar Polar Orbiter (LPO) was put forward. Three times it was proposed for funding. Twice it was shot down by the Office of Management and Budget; the third time it never even made it out of NASA. (Andrew Chaikin, "A Guided Tour of the Moon," *Sky and Telescope*, 68,

September 1984, 211). LPO was put on hold. LPO was so low in priority that it did not make the list of core objectives NASA formulated for its program of unmanned space exploration through the year 2000. Meanwhile, the USSR began talking about flying a vehicle like LPO. United States scientists even met with their Soviet colleagues to discuss putting American instruments on the Russian space probe. The Soviets' major incentive in flying the mission seemed to be that the United States was not. ("U.S. Sensors May Go on Soviet Moon Mission," *Science News,* 113, June 3, 1978, 358).

As part of a recent resurgence of interest in the Moon, NASA has again opened the books on a Moon-orbiting spacecraft. In its third incarnation, it is the Lunar Geoscience Orbiter (LGO). The time for the LGO has come. It is appropriate that while we obtain pictures of the satellites of Saturn and Uranus we also initiate a modest but thorough investigation of our own sister world. We can do this in the 1990s with a sophistication that was not possible in the 1960s.

In many ways, the LGO will resemble one of the several Earth-resource satellites operating today around our own planet. As its name implies, its emphasis will be geochemical. The LGO will be able to spot concentrations of chemical elements and magnetic ores. Another goal may be cartographical. Our current mapping resolution on the Moon is 1 kilometer on the Nearside and 10 kilometers on the Farside. The LGO could improve this by factors of 10 and 100 respectively. (Erwin J. Bulban, "Economic Benefits of Lunar Base Cited", *Aviation Week and Space Technology,* 118, April 18, 1983, 132.)

The LGO will be put into space by the Space Shuttle. It may fly 6- to 12-month sorties in the vicinity of the Moon, after which it will be returned to a near-Earth orbiting **space station** for refurbishment and transfer of data. It could then be sent to the Moon again—or to Mars.

THE RETURN TO THE MOON

Even before Apollo XVII left there, the call had gone out for a return to the Moon. In the heyday of the Apollo program, its lunar landings were considered to be only precursors to a more ambitious exploration program, a second generation that would pick up where Apollo left off without losing a step (Figure 9–7).

Two months after Apollo XI, NASA published *America's Next Decade in Space,* which outlined an ambitious extension of lunar exploration. This project, at one time called the Independent Lunar Surface Sortie (ILSS), centered around a **moon base** to house crews on the Moon for extended periods—two months to a year. The ILSS called for a manned polar-orbiting space station in lunar orbit that would house up to eight people. This station would send out three- or four-person teams on two- to four-week sorties to the lunar surface aboard a LM-like "taxi." Finally, components for a permanent 12-person base would be sent down and assembled on the lunar surface. (Daved Dooling, "An-

9–7 The last men to visit the Moon depart at 5:54:36 P.M. EST, December 14, 1972.

(NASA photo)

other Decade May Elapse Before Man Returns to the Moon," *Space World,* pp. 9–191, November 1979, 24).

The ILSS would have required placing payloads of 4,500 or more kilograms on the Moon—a weight greater than that of all the Apollo payloads combined. Using existing throwaway technology, the lavish project would have been prohibitively expensive. Eight Saturn V launches would have been required to establish the base alone!

A moon base was impractical in the 1970s. However, with the advent of the reusable Space Shuttle, it has become a possibility again (Figure 9–8).

Two weeks after and 10,000 kilometers away from the site of the Conference on the Origin of the Moon, another Moon meeting took place in Washington, D.C., October 29 to 31, 1984. This symposium was held at the National Academy of Sciences and was titled "Lunar Bases and Space Activities of the 21st Century." Scientists and engineers gathered to define the likely trends in space exploration after the establishment of the United States' permanent spacestation in the 1990s. Their objective was to identify programs for technology development that are needed now in order to meet the goals of the next several decades. From the symposium title, it is immediately clear that one of these goals may be the establishment of a permanent manned facility on the Moon. Just as the LGO marked revived interest in selenology for the 1990s, this meeting indicated that serious consideration is again being given to returning people to the Moon.

One of those at NASA who maintained a serious interest in a manned lunar presence during this period was Michael Duke. Before the Washington symposium, Duke chaired a meeting of a "Lunar Base Working Group" (LBWG), formed to study the issues involved in establishing a moonbase. Fifty people met in Los Alamos, New Mexico, in April 1984 to define the scientific and technological studies that are needed so that an intelligent decision can be made on whether to carry the manned space program to the Moon again. It was agreed that the time for making this decision

(NASA photo)

9–8 Space Shuttle Orbiter Challenger begins its maiden voyage on April 4, 1983.

may only be less than 10 years off. This working group concluded by issuing a white paper that called upon NASA to adopt a lunar base as an official goal.

Another working group met in La Jolla, California, to discuss the social and political ramifications of such an endeavor. Reports of the two working groups were reviewed at the symposium, and papers were read on subjects ranging from "Integrated Ecological/Biological and Engineering Principles Applied to an Ecological Life Support System

(CELSS) on the Moon" to "Lunar Bases and Extraterrestrial Law: General Legal Principles and a Particular Regime Proposal."

Why this interest now? After all, the idea of a return to the Moon has been tossed around before. It may seem unreasonable to talk about lunar ventures once again after a decade of cuts in NASA's budget for planetary exploration. There are several reasons, though, why the time is indeed right.

First, there is the Space Shuttle, which has singularly reduced the cost of carrying materials into and out of space. Another is presidential support for a space station called the Space Platform. This facility has been tentatively approved by Congress, and installation of hardware in space could begin as early as 1992. The station would form an embarkation platform for further lunar exploration. The LBWG recommended that the Space Platform be configured in anticipation of a possible role in supporting a lunar base. Using the appropriate architecture now would save costly modifications later on (Figure 9–9).

Other factors favoring a moonbase at this time include the currently perceived favorable status of the economy. Also, private industry has shown increased interest in getting into space. There is a possibility for international cooperation on building the moonbase. Other countries in addition to the United States may participate in the Space Platform effort. Last, there seems to be a more positive public attitude toward space exploration now than there was a decade ago.

Planners are optimistic. People could be at work on the project by the end of the century if the Space Platform stays on schedule. The moonbase could become a reality by the fiftieth anniversary of the first artificial satellite in 2007. The gap between Apollo and the moonbase may not be unusual. NASA Deputy Director Hans Mark has drawn a parallel between it and the 31-year interval between Roald Amundsen's arrival at the South Pole and the construction of the first permanent station there.

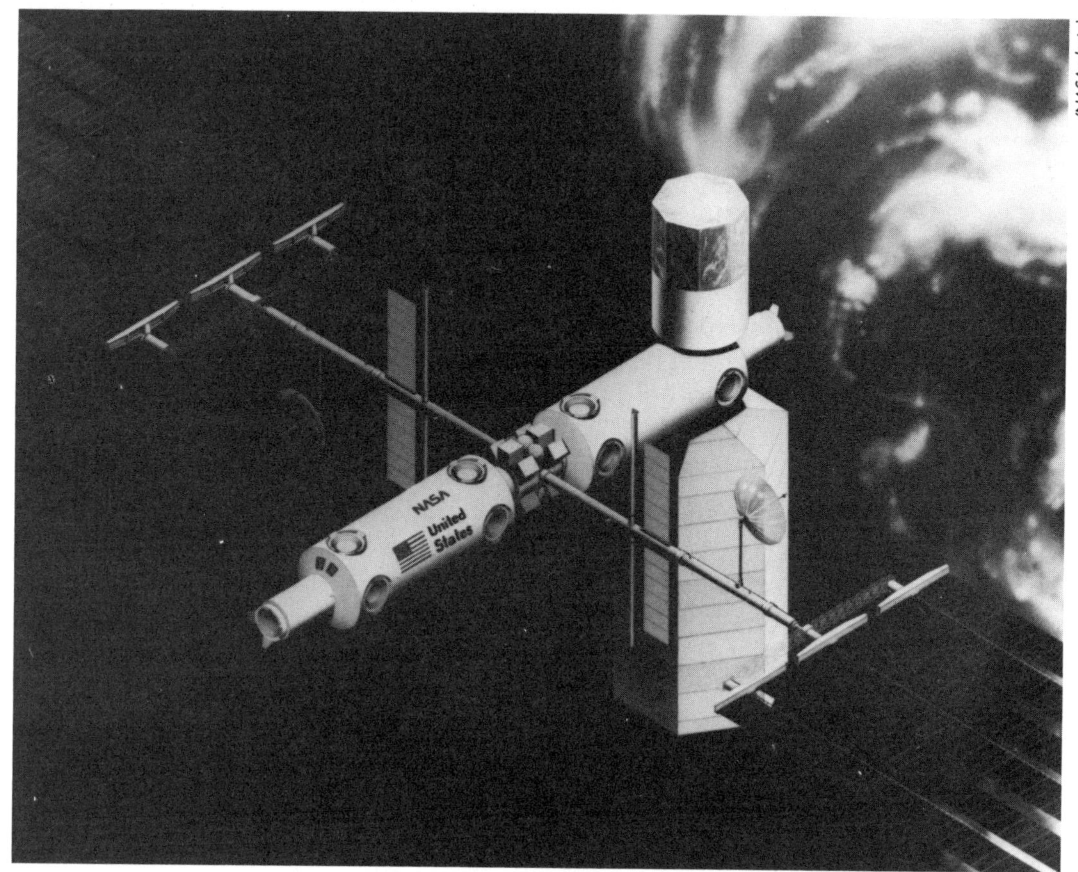

(NASA photo)

9–9 A possible spacestation configuration. The components would be designed to fit into the Space Shuttle Orbiter cargo bay. Additional components could be added as needed. A design study for the station began in 1982.

One might ask: How much will it all cost? As a point of comparison, from 1962–1972 NASA received 0.3% of the Gross National Product (GNP) for the Apollo program. The moonbase project would be spread out over an interval at least twice as long as that of Apollo, however. Also, the United States' GNP has increased by a factor of two since the 1960s. It is possible that the project could be undertaken for less than 0.1% of the GNP. NASA could support it on its existing budget if the operation costs of the Space Shuttle continue to be reduced. The key to the plan is that the project receive *sustained* funding over a long period and not "rollercoasting" allocations that have plagued other federal projects. Obtaining funding may not be as difficult as will be the development of long-range government planning (in terms of decades). Such planning will be necessary in any case as the pace of development of our technological society continues to accelerate.

Cost savings may be achieved by inviting international participation. Also, if the Space Platform were to be viewed as a *prototype* for the moonbase (many of the assigned tasks will be similar), research and development costs would be reduced.

Major funding is in the future, of course. Right now, the LBWG proposes only that the LGO program (viewed in the role of an early reconnaissance for the base) be continued and that a modest pilot program be initiated to research the use of lunar materials.

Actual work on the lunar base could begin as soon as the Space Platform was completed. By this time a larger Space Shuttle would be necessary. This Shuttle-Derived Launch Vehicle (SDLV) would be capable of putting significantly larger payloads into Earth orbit. With room for increased cargo, the SDLV would bring the current cost of $2,000 per kilogram down.

Once the components for the moon base were put into Earth orbit, the hard part would be over. Still, they would need to be ferried to the Moon. Concurrent with Space Platform development would be the design and building of an Orbital Transfer Vehicle (OTV). The purpose of the OTV would be to transfer payloads from low orbits such as that of the Space Platform (around 150 kilometers up) to **geosynchronous orbit** (35,900 kilometers).

Geosynchronous orbit is the distance from the Earth at which an orbiting spacecraft will revolve about the Earth at the same rate that the planet turns. The satellite or other payload situated there appears stationary with respect to any point on the Earth's surface. For this reason, it is strategic for communications and defense-related work. The energy necessary to reach geosynchronous orbit is only slightly less than that required to go to the Moon. The OTV could easily be modified to take men and material from low Earth orbit to lunar orbit.

The OTV should be available sometime shortly after the Space Platform to replace the "strap on" rockets now being introduced to send payloads to geosynchronous orbit from the Space Shuttle cargo bay. The OTV would be a true space craft, designed to function only in space. Therefore, it would not need any of the sophisticated aeronautical engineering that allows the Space Shuttle to operate in the atmosphere. Initially, the OTV would be automated; eventually it would be manned.

After the OTV placed an object in geosynchronous orbit, it would return to the Space Platform by using a technique called **aerobraking.** (A method of aerobraking was vividly portrayed in the 1984 movie, *2010: Odyssey Two*). As the OTV swung back to the Earth on its elliptical path, it would skim the top of the atmosphere. Friction would relieve the spacecraft of energy, and it would emerge in a low Earth orbit. An OTV returning from the Moon would operate in the same way.

In its role as lunar barge, the OTV would probably have to be able to deliver a cargo weighing 20,000 kilograms, equal to the mass of a Space Platform module. Thus, the components of the moonbase would arrive in 20-metric-ton lots and would in some ways resemble those of the spacestation.

Near the Moon, large lunar landers (based on the technology developed for the Apollo LM) would transfer cargo the rest of the way to the lunar surface. Some of these landers would be reusable. Eventually, it would be desirable to put a manned space station in orbit around the Moon to serve as a staging area for operations on the ground.

The evolution of the actual moon base might follow a scenario drawn up at the NASA Johnson Space Center. The precursor to people returning to the Moon may be extensive automated exploration. Remotely controlled rover vehicles such as the Russian Lunokhod might be used for this purpose. Once a site was chosen, a temporary or semi-permanent manned scientific outpost would be established much like the ones in Antarctica. Eventually, use of lunar resources would begin, much of it automated. Finally, as the Moon began partially to pay for itself, a permanent base for fundamental scientific research would be built. In time this base would become self-sufficient and nearly independent of the Earth.

Where might the moon base be established? Probably it would be built at the site of one of the Apollo landings because we know these places best. Expeditions could then be sent out from here to points of interest. They might even establish temporary shelters at other sites that could be visited intermittently by people from the main base.

Two unexplored areas also hold promise: the lunar poles. In these regions where the Sun is never high in the sky, there may be crater basins and other shaded spots where it never shines. (The interiors of craters Peary and Amundsen are examples.) It is estimated that this condition may exist over 1/2 to 2% of the lunar surface. In these permanently dark recesses, the temperature would remain around −100° C. constantly. It is possible that frozen patches of water still cling to the Moon in such places. Such water could exist as permafrost just under the lunar soil. Discovery of even modest amounts of water on the Moon would be a boon to the moonbase and reason enough for putting its location in close proximity to this vital resource.

Although there has been no trace of water on the Moon found to date, the LGO will make a concerted effort to find it as that spacecraft flies over the poles. One way it might do so would be to detect ice as a gravity anomaly. (Thornton Page, "Apollo Remembered: Conference Notes from Houston," *Sky and Telescope,* July 1979, p. 28).

The actual moonbase would need to be protected from the viscious lunar environment. An early proposal suggested using nuclear explosives to blast habitable caverns out of the moonrock! It is more likely that a building on the Moon may be a modified Space Platform module buried in lunar soil. Such a habitation could house six to 12 people. It could be installed by automatic landers and remotely controlled earth movers before its occupants even got there.

A covered or underground moonbase is necessary to protect its inhabitants from dangers raining down on them from space. A meteroid hit, unhampered by an atmosphere, could be catastrophic. The odds of a large body just happening to penetrate the moonbase are miniscule, though. Much more dangerous is the continual presence of radiation from space that bathes the Moon constantly. A solar flare could temporarily raise lunar radiation above the level at which a spacesuit or thin spacecraft wall would be able to shield an astronaut. The Apollo LM was vulnerable to such radiation (James Michener wrote about the effect of a solar flare on an Apollo Moon mission in his 1982 novel, *Space*.) Dirt is an efficient and easily obtainable radiation shield. For this reason, lunar bases may resemble large fallout shelters.

Another benefit of soil is its effectiveness as a thermal insulator. A blanket of lunar soil would simplify heating and cooling requirements in the base during the extremes of the lunar night and day. Eventually lunar bases will be underground at a level at which the ambient temperature remains consistent.

Science fiction has imagined capping an underground lunar city with a large transparent dome filled with air. Such a dome would have important aesthetic benefits in making life on the Moon more like that on the Earth, but it would have to be shielded from radiation. Perhaps great louvers outside the dome could regulate solar heating like giant Venetian blinds.

Transportation will be important in all future activities on the Moon. Mobile tenders will be needed to unload lunar landers. Landmovers will be required for construction. Mooncars with ranges exceeding that of the LRV will be necessary to expand the range of transportation.

Long- and medium-range transportation will probably remain powered by rockets. A science fiction staple is the rocket belt. Such a device may come into vogue for personal travel requirements on the Moon.

The low lunar gravity will allow some transportation experiments on the Moon that are not possible on the Earth. In the 1960s, a computer-controlled, gyroscopically stabilized pogo stick was suggested for use by astronauts on the lunar surface! Such a device could transport a person in a series of 15-meter hops at the speed of 11 kilometers an hour.

For short-range transportation, walking—in the loping style of the Apollo astronauts—will still be used a lot. Under lunar gravity, an individual weighing 60 kilograms will weigh only 10 kilograms. Certainly, much of this weight is made up for by the heavy moonsuits and life-support equipment an astronaut must carry along on the lunar surface. This paraphernalia will become lighter, however. For instance, the bulky inflated suit will give way to a skin-tight elastic suit that will maintain pressure in the body physically rather than by pressurization.

With lighter accouterments, an astronaut will be more comfortable during extended stays outside the moonbase. She or he will also be able to pack heavy scientific loads. An unencumbered astronaut on the Moon will be able to leap 3 meters into the sky and bound from place to place in giant strides. An ever-present danger will be the tendency to overestimate what can be done on the Moon. The astronaut may not weigh as much on the Moon as on the Earth, but his or her inertia will be the same. That is, the astronaut will be able to achieve a greater speed but will have the same difficulty in stopping. There may be a good many broken bones to set at the first lunar base!

The moonbase will always need electrical power. Solar power generators will be more efficient on the Moon than on the Earth. The lack of an atmosphere will always allow all the energy of the Sun to fall continously and unobscured on solar collectors throughout the two-week lunar day. However, batteries will have to store enough power to supply needs during the two-week night.

During the night, fuel cells or a nuclear power plant might be used. The contrast in temperature between day and night may itself be put to work generating electricity. It is likely that power on the Moon will come from a mixture of sources in order to provide a reliable and steady supply.

Because there is no air on the Moon, communications cannot depend on sound (though two astronauts' helmets pressed together could conduct a voice). Reliance will be on radio signals, (although a series of signal mirrors on the Moon has been proposed.) Without an atmosphere to bend them, however, radio communications will be limited to line of sight. This will constrain long distance communications on the Moon where the horizon is only half as far away as it is on the Earth. Signals can always be relayed by means of the ever-present Earth (on the Nearside), but this would be inefficient and a nuisance because of the slight time delay caused by the finite travel time of the radio waves. More likely, a series of communications satellites will be put into lunar orbit for this purpose. These satellites could even be linked into one large communications system by long thin wires.

Basic to human needs on the Moon—as on the Earth—will be air, food, and water. We have already talked about water. Air is one subject of our discussion of lunar resources in the next section. For a moonbase ever to approach self-sufficiency, it will have to produce its own food. Initially, efficient organisms such as algae may be grown to produce nutrients. This may be replaced by hydroponic farming. Inefficient sources of food such as livestock will have to wait a while.

Could crops ever be grown on the Moon itself? By inserting organic material such as microorganisms into the moon soil, it might be possible to make it fertile. One can envision large lunar farms in airtight enclosures on the lunar surface. They would look like hothouses do on the Earth. To conserve air, the farms could be kept in a high-oxygen, low-pressure atmosphere. These enclosures would also have to control temperature and light. Plants would have to be chosen or forced to adapt to some unusual conditions. For instance, what will be the effect on a plant that grows while it is exposed to sunlight of the two-week-long day? How might the lunar gravity, which is only a sixth of that of Earth's, affect a sunflower?

It might be possible one day to genetically engineer food plants to survive well on the Moon. It has even been suggested that cactus-like plants might be created that would live outside on the bare lunar surface itself. Such plants would release oxygen into the primitive lunar atmosphere. This would be the first step toward **terraforming** the Moon—modifying it to become more like the Earth. Such a practice would take a long time, and starting it is far in the future. Whether it would even be right to do so is an intriguing problem for ethical futurists.

In the immediate future, the Moon will remain an alien environment. For people to survive there, they will have to adapt to its ways and scrupulously take care of its meager resources. All commodities necessary for life will have to be efficiently recycled before the moonbase can survive on its own. If it succeeds in doing this, though, the Earth's societies might learn a thing or two from their lunar counterpart. As this point approached, lunar settlement would become lunar colonization, and the entire Earth–Moon system would become the home of humankind.

THE FUTURE

Finally we come to the question: Why return to the Moon at all?

Scientifically speaking, the answer is: for some of the same reasons that we went there the first time. Many of the questions that Apollo set out to answer weren't but not because Apollo was a failure. The Moon was supposed to have a geologically virgin, relatively simple surface. Its record was supposed to be laid out for all to see. This was not the case. In fact, it was found that the Moon's surface was complex and that endogenic and impact activity had effectively blotted out its earliest history even in the highlands.

We need to return to the Moon to study its gross geologic and physical characteristics. Apollo was a spot check on these properties. However, on a global scale, the Apollo landing sites were isolated, all being clustered about the equator on the Nearside and in relatively smooth areas (Figure 9–10).

By analogy, only consider what would happen if a dozen space probes were sent to explore the Earth randomly. For one thing, chances are that most of them would land in the ocean. Also, a probe would give a skewed impression of the Earth's surface if it happened to land in, say, a supermarket parking lot.

9–10 Sites from which lunar samples have been retrieved (A = Apollo, L = Luna).

(Courtesy of New Mexico State University)

To probe the lunar interior, a network of at least 30 seismic stations needs to be established over the entire surface. Such a network would chart the way seismic waves travel through every part of the Moon and should be left in operation long enough to respond to rare large impacts or be actively triggered by artificial impacts. In addition, thermal probes need to be sunk into the Moon at various points to measure heat flow from the interior and to detect inhomogeneities. Only two such probes were deployed during Apollo.

Much of the further exploration of the Moon can be done by using automated spacecraft. Orbiters can geochemically and magnetically map the Moon. Landers can return samples to the Earth. Alan Binder has advocated a program of exploration called Selene, which would consist of 34 lunar landers. Eighteen of these would return samples. (Selene would require 14 Space Shuttle flights.)

Could a global survey of the Moon be conducted by unmanned landers alone? They would certainly be cheaper and provide more data points for the money. The technology for building such space probes has improved considerably since we last set foot on the Moon. Ultimately, though, people will have to be sent. Only a person trained for the task can adaquately choose this rock or that one as an important sample or react appropriately to unexpected conditions. Too, all automatic devices break down from time to time. If human beings make them, human beings will have to go fix them occasionally. Astronauts need to return to the Moon if only as repairmen! Their ability to do so and to successfully complete missions that would have been thwarted by malfunction if under machine control has been demonstrated time and again.

Remotely controlled operations *will* become important on the Moon. The human brain cannot be duplicated but the human arm and back can be. Still, it will always be necessary now and then for a woman or man to take a wrench to a valve or to secure a scientific instrument or an intriguing specimen that's just out of the reach of robot hands.

Of what use is scientific information about the Moon, anyway? Besides its aesthetic worth, there could be unanticipated benefits, as is always the case with fundamental research. One thing we do know: Whatever happened early in the Moon's history probably happened to the Earth as well. At some point the evolution of these two worlds diverged, though. Knowing how the Moon is the same and how it is different from the Earth will tell us about planets in general and our own world in particular.

Scientific advance is but one tiny aspect of human affairs. People undeniably have an intrinsic thirst for knowledge; it is not necessary to justify science in this day and age. Yet its relative importance on the scale of human needs is still debated, especially when the price tag appears.

The Apollo program was not exclusively or even primarily a scientific undertaking. If it had been, it would never have gotten off the ground. Political considerations were paramount. This will continue to be the case in the future. The first lunar base will be prestigious for the country that establishes it. It will be a source of pride and will occupy an auspicious place in human history. It will have technological spin-off benefits. It will offer political and strategic assets.

The Moon may also have commercial value (In 1984 a Texas savings and loan association sought approval from state banking authorities for opening a branch office on the Moon!) One of the goals of LGO will be to look for profitable lunar resources such as metal ores that might one day be mined on the Moon. As early as 1965, the U.S. Bureau of Mines undertook a study of the potential for lunar mining.

Lunar resources need not be exotic materials to be useful. If it becomes possible to process oxygen out of moon soil (water would be a by-product), liquid oxygen could be used as a propellant for lunar landers and the OTV. It could even be used to refuel Space Shuttle orbiters. Why export oxygen from the Moon to use in space when it is plentiful and easy to get on the Earth? The reason is that it takes only

$1/20$ the energy to launch cargo from the Moon as it does to haul it into space from the Earth. Thus, lunar-derived materials might actually be cheaper, even for use near the Earth.

Just plain old lunar dirt may be an exportable resource. Future Earth-orbiting space stations (of particular interest to the military) will be in polar orbit. Twice on each revolution such a station will be exposed to great quantities of cosmic and solar radiation near the poles. (The Earth's magnetic field channels charged particles toward the poles.) Thus, these stations will need extensive radiation protection. Layers of lunar dirt would constitute a cheap and effective radiation shield. The ablative heat shields of aerobraking OTVs could also be made out of moon dirt.

Useful metals are ample on the Moon. Iron, aluminum, titanium, and silicon could be freed from moonrock by chemical and electrochemical methods and then be used there or sent to the Earth. Ilmenite, an oxide of iron and titanium, is found frequently in concentrations on the Moon that would be considered ore-quality on the Earth. It could be melted down in a solar furnace (a heat source that is available continuously for two cloudless weeks each month) and then separated out electrostatically. Meteorites are another source of iron on the Moon. Mining will be a good activity on the Moon to enable the lunar base to pay its way. Mineral resources will be an important factor in deciding where to put the base.

After mining, the next major industry to arrive on the Moon will be manufacturing. For a lunar colony to flourish, it must learn to use local building materials. Construction using Moon-fabricated ceramics might also be of use in Earth orbit. Currently, no research is being conducted anywhere on the use of intrinsic lunar materials for manufacturing. One of the proposals of the LBWG was that work be undertaken in this area.

The lunar vacuum will aid in the production of pharmaceuticals and electronic components that might be contaminated in a less-than-ultraclean environment. Pure metals could be produced there, and any vacuum coating process would be much easier to perform.

The low lunar gravity would also be useful in making optical instruments and intricate mechanical parts. Totally new products might be originated and made on the Moon as well. One day the label "Made on the Moon" on a great many lunar exports will stand for extreme purity and finer precision.

Although a lunar colony may never be independent of the Earth for some of its needs, it is likely that it and the Earth will become interdependent, a process that will begin when lunar ferries transporting supplies to the Moon begin carrying lunar-derived cargos back to the Earth on their return trips.

It may be possible for there to be one-way Moon–Earth traffic as well. The Moon's low escape velocity (2.38 kilometers per second) makes it considerably easier to fling a mass to the Earth from the Moon than vice versa. (The escape velocity for the Earth is 11.19 kilometers per second.) Physicist Gerard O'Neill has proposed a method for throwing objects off of the Moon with the use of a **mass driver** (Figure 9–11). This device would consist of a long track across the lunar surface. Cargo-carrying units would be magnetically suspended above the track. In the lunar vacuum, this system would be almost frictionless. It would be possible to accelerate the cargo units electrically to tremendous speeds. When it left the track, a cargo unit could be traveling at escape velocity. It need have only small thruster rockets for steering on the rest of its journey to the Earth. Eventually such a device could be manned.

In addition to facilitating practical industry, the Moon provides a unique laboratory for fundamental scientific research that requires the conditions that exist there.

Cryogenic physics will be easy during the long lunar night. Specifically, research into **superconductivity,** the loss of electrical resistance in materials at extremely low temperatures, will continue and accelerate the miniaturization of electronics.

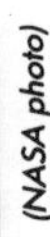

9–11 A future mining "town" on the Moon. In the center of the picture, transport vehicles prepare to lift lunar resources to orbit. To the right, the cylindrical modules with large windows are agricultural facilities for growing food. Living quarters are behind them, nestled into the hillside. A mass driver runs along the left side of the picture.

In the lunar vacuum, subatomic particle beams and accelerators can operate without elaborate pressure housings. Experiments will be performed there free of the background "noise" of particles produced by cosmic rays in the Earth's atmosphere. This environment is so perfect that physicists who will flock to it may one day demand an end to the "pollution" caused by the formation of a tenuous lunar atmosphere of rocket engine exhaust gases from arriving and departing spacecraft! Mass drivers will reduce some of this potential problem, it is hoped.

Biologists will be able to study the effects of long-term exposure of reduced gravity on the Moon. They will be able to analyze the effects on living creatures, including people, of the absence of a magnetic field.

It is perhaps fitting that the science to benefit the most from a lunar base will be astronomy. Beer and von

Mädler first noted the advantages of studying astronomy from the Moon. More recently, a later Apollo experiment that was never funded would have established a telescope with a 25-centimeter aperture on the Moon that would have been remotely controlled from the Earth.

On the Moon, there is no atmosphere to obscure or blur an image. Stars are visible in the daytime. Observations could be made very near the Sun. At night (all two weeks of it) the sky is darker than at any terrestrial observatory. A Moon-based telescope could look at wavelengths of light that never penetrate our atmosphere at all: the ultraviolet and infrared.

Because of the lunar vacuum, a telescope would not have to be shielded from weather. Because it could be efficiently insulated, it would not be susceptible to thermally induced stress, which can deform a telescope and misalign the instrument's optical path.

In the low lunar gravity, telescope mirrors could be made much larger than on the Earth. Gravity would deform them only a sixth as much. Because of this, refracting telescopes might stage a comeback on the Moon.

Lunar astronomers would never have to climb above the atmosphere to a mountaintop in order to observe and would never be "clouded out." Of course, with options for remote operation and data collection, they might not need to be at the telescope at all. The astronomers could even be in their living rooms on the Earth!

Many of the advantages of lunar-based astronomy also hold true for a telescope in space—for instance, the earth-orbiting Hubble Space Telescope. However, because such a telescope falls around the Earth, it must continuously move to track the object it is looking at; frequently the Earth gets in the way. The lunar telescope has an advantage over its nearer-to-Earth counterpart for observing projects that need prolonged, continuous periods of focusing on a single object. (Remember that objects rise and set much more slowly in the lunar sky.) The lunar-based telescope is also advantageous because it provides a steady mount for making delicate and precise positional measurements.

Presently, the mirror of an orbiting telescope is limited in size to the diameter of the Space Shuttle's cargo bay. A mirror of nearly any size could be made on-site anywhere on the Moon. There is certainly enough silicate to produce lunar glass. There are even readily available mirror mounts waiting on the Moon—the bowl-shaped lunar craters!

A telescope on the Moon would probably be used for high-resolution projects such as examining distant comets or searching for planets orbiting other stars. Another use would be looking at distant objects such as the remnants of supernovae.

In the longer wavelengths, radio astronomers would listen to radio-frequency emissions from space with huge antennas erected on the Farside. With the high resolution that these dishes could achieve and the stark, radio-quiet environment that this hemisphere provides (shielded from all signals emanating from the Earth by the Moon itself), radio astronomers could listen in on the universe, perhaps searching for signals produced by civilizations elsewhere.

Astrophysicists who study high energy particles traveling through space would have an observation post on the Moon that would be outside the Van Allen belts. They could even study the history of this radiation by looking backward. In the lunar soil lies the cosmic ray record implanted in the lunar surface over the eons. They need only to dig back through time.

Much of what has been said about activities on the Moon in the age of the moonbase has been speculative. Extrapolating further into the future is really a flight of fancy but one we should indulge in as we conclude.

Let us now imagine a time when the Moon has been colonized. Women and men are born and die in lunar cities, perhaps never visiting their ancestral home the Earth. Is such a vision a dream or a nightmare? Why would anyone want to live out life on the Moon? Despite the hazardous conditions there, there will be those with a frontier spirit ready to face them, and once a lunar city is established, life there will have several distinct advantages over existence on the Earth. The lunar gravity will be healthful to those people with heart problems and other ailments caused by physical stress on the body. Also, there are no lightning storms, tornadoes, hurricanes, blizzards, or floods on the Moon to interfere with human life and property.

To this list of nonexistent disasters on the Moon we can, one hopes, add war. In fact, the Moon has the potential to become a paragon of human cooperation. Life on the Moon will have to be in harmony with its environment *vis-à-vis* the recycling and care of scant resources. Just as necessary will be harmony among the inhabitants, those persons from different nationalities and backgrounds who will join together to become lunar citizens. This is not idealistic optimism; it will be essential to life on the Moon. There are too many strikes against us there already without introducing distrust and animosity to hinder our success.

What then? After humankind has had practice adapting to a new world, it will be ready for others. Skills learned in using lunar materials can be used on Mars or in the asteroid belt. Indeed, the Moon is an excellent embarkation point for exploring the other planets and ultimately the stars.

We have dealt so far with how people will affect the Moon to make a home out of it. How will the Moon affect people? Will the lunar environment change them? One can imagine that the day/night cycle implanted in humans might be modified on the Moon. Magnetic deprivation may have its effect as well. The psychological alterations can only be guessed at. Will people of the future adapt to life on the Moon? After multiple generations, will they evolve into a new species, *homo lunae*?

Much of this future history of the Moon may never happen. It can be thwarted or altered by innumerable circumstances. However, if humankind survives as a race, its expansion into the universe seems destined. It also seems likely that men and women of centuries to come will remember the Moon as the first step out of the Earth's cradle. I would like to think that someone, somewhere reading this book will one day visit the Moon. Let me know when you get there! That is, of course, assuming that I don't make it there first. . . .

GLOSSARY

accretional heating. Heating of a body caused by the kinetic energy of smaller particles slamming into it ("accreting").

accretion disk. A ring of particles around an object that is coalescing into a satellite.

aerobraking. Using the friction of an atmosphere to modify the orbit of a spacecraft.

albedo. The percentage of incident light that a body reflects.

alpha particles. Positively charged particles, consisting of two protons and two neutrons, often released in nuclear reactions.

annular eclipse. An eclipse of the Sun that takes place when the Moon is near apogee so that its disk doesn't quite cover the disk of the Sun.

anomalistic month. The period of revolution of the Moon when measured with respect to its Line of Apsides (27.56 days).

anorthositic gabbro. A dark, plutonic rock consisting mostly of coarse crystals of calcium plagioclase.

aphelion. The point in the orbit of the Earth at which it is farthest from the Sun.

apogee. The point in the orbit of an Earth satellite at which it is farthest from the Earth.

arcuate rilles. Gently curved crevasses on the Moon.

ascending node. For the Moon, the point at which its orbit crosses the ecliptic as the Moon moves from South to North.

astrolabe. An instrument used to determine the altitude of a celestial object.

Baily's beads. Points of light seen around the limb of the Moon just at the moment of totality in a solar eclipse. They are caused by light streaming through valleys on the Moon.

barycenter. The center of mass of a system of objects.

basalt. A dark-colored igneous rock composed primarily of feldspar.

breccias. Rocks composed of sharp angular fragments cemented in a fine-grained matrix.

caldera. A large crater formed by the collapse of a volcanic cone.

capture (theory of the origin of the Moon). The suggestion that the Moon was formed elsewhere in the solar system and came near enough to the Earth to be captured into orbit by the Earth's gravity.

Cassegrain. A type of reflecting telescope in which the light from the secondary mirror is directed to the observer through a hole cut in the center of the primary mirror.

central lunar eclipse. A lunar eclipse in which the Moon passes directly through the center of the Earth's shadow.

chromatic aberration. A smearing of an image produced by the fact that a lens will refract different colors of light by different amounts.

chronometer. An accurate clock.

coaccretion (theory of the origin of the Moon). The suggestion that the Earth and Moon formed as two bodies from the same original cloud of material.

constellations. A particular configuration of stars that stands out and has been given a name; also, the region of the sky where those stars are found.

core. The central part of a planet or satellite.

cosmic rays. Charged particles from space.

cosmogenic. Formed in or coming from outer space.

crater counting. Counting the number of craters visible in a region to determine the age of the surface there.

craters. More or less circular depressions formed by volcanic explosions or the impacts of meteoroids.

crust. The outermost solid layer of a planet or satellite.

deceleration dunes. Concentric rings around a crater caused by the collapse of a rising cloud of material.

deferent. In the Ptolemaic system, a stationary circle along which an epicycle moves.

descending node. For the Moon, the point at which its orbit crosses the ecliptic as the Moon moves from North to South.

docking. Coupling two spacecraft in space.

draconic month. The period of revolution of the Moon when measured with respect to its Line of Nodes (27.21 days).

earthshine. The faint illumination of the Moon by light reflected off the Earth that allows one to see more of the Moon's disk than is actually being illuminated by the Sun (also called "ashen light").

eccentricity (of an ellipse). The ratio of the center-focus distance to the semimajor axis; a measure of how "squashed" an ellipse is.

eclipse. The partial or total obscuration of one celestial body by another, as when the Earth or Moon moves into the other's shadow.

ecliptic. The plane of the Earth's orbit about the Sun; equivalently, the apparent path of the Sun's motion on the celestial sphere.

ejecta. Material ejected from a crater during its formation.

electromagnetic radiation. The whole array of electromagnetic waves, from gamma rays through visible light to radio wavelengths.

ellipse. A curve formed by the intersection of a circular cone and a plane cutting through the cone.

endogenic. Formed by internal processes.

epact. The 11-day difference between the lunar and solar year.

epicycle. The circular orbit of a "planet" in the Ptolemaic system, the center of which revolved around another circle (the deferent).

equant. In the Ptolemaic system, the point at the center of a deferent (not necessarily the Earth).

escape velocity. That velocity an object must reach in order to escape from the gravitational attraction of another body.

evection. The periodic change in the eccentricity of the Moon's orbit caused by the perturbation of the Sun.

exogenic. Formed by external processes.

eyepiece. A magnifying lens with which one views the image produced by the objective of a telescope.

faults. Discontinuities in a rock formation caused by a shifting of the crust on one side relative to the other.

fines. Lunar "soil," defined as particles with diameters less than 1 centimeter.

fire fountaining. Low-viscosity lava spewing up violently.

First Quarter. The point 7½ days into the lunation when we see exactly half (the west half) of the Moon's disk illuminated.

fission (theory of the origin of the Moon). The suggestion that the Earth and Moon were originally one body and that the Moon was torn from the Earth by some cataclysmic event.

focal length. The distance from a lens or mirror to the point at which most of the light is brought to a focus.

focal plane. The area over which the light from a telescope is in focus.

focal point. The point of the focal plane that is on the axis of symmetry of the instrument.

focus, focii (of an ellipse). Points associated with a conic section such that the distance from either focus to any point on the curve is the same as that from that point to a straight line called the directrix.

fuel cells. Electrochemical cells in which the reactants (often hydrogen and oxygen) are stored separately and supplied to the electrodes on demand.

Full Moon. That phase of the Moon halfway through the lunation in which we see the whole disk of the Moon illuminated.

geocentric. Centered on the Earth, as in the concept that the Earth is the center of the universe.

geosynchronous orbit. An orbit in which a satellite revolves around the Earth at the same rate as the Earth rotates and therefore stays at the same spot in the sky all the time.

ghost craters. Craters that have been almost entirely covered by lava flows occurring subsequent to their formation.

graben faults. A discontinuity in the surface of a body caused by a block of crust dropping.

gravitation. That force of attraction which bodies with mass have for each other.

Harvest Moon. The Full Moon nearest to the autumnal equinox when the Moon's retardation is at a minimum and a nearly Full Moon is visible in the evening sky for several consecutive days.

heliacal. The rising of a star when it is first visible emerging from the glare of the Sun just before sunrise.

heliocentric. Centered on the Sun. (In the context of this text, the idea that the Earth and other planets revolve around the Sun.)

highlands (lunar). The older, more heavily cratered areas of the Moon's surface; also called *terrae.*

Hunter's Moon. The first Full Moon following the Harvest Moon.

impact (theory of the origin of the Moon). The suggestion that the Moon was formed out of the debris created in a collision of the Earth with one or more differentiated planetesimals.

inclination (of an orbit). The angle between the orbital plane of an object and some fundamental reference plane (e.g., the angle between the plane of the Moon's orbit and the ecliptic).

intercalation. The adding of extra days or months to one calendar system to make it conform to another.

Kepler's Law's of Planetary Motion. Three laws, formulated by Kepler, that govern planetary motion:

1. The orbits of the planets are ellipses.
2. The line connecting a planet and the Sun will sweep out equal areas in equal amounts of time.
3. The square of the period of a planet's orbit is proportional to the cube of the length of the semimajor axis of the orbit.

kinetic energy. Energy associated with the motion of an object.

KREEP. An acronym for lunar rocks rich in potassium (K), rare-earth elements (REE), and phosphorus (P).

last quarter. That phase of the Moon three-quarters of the way through the lunation when we see the east half of the Moon's disk illuminated.

launch escape system. A rocket on top of a spacecraft designed to pull it and the crew clear of danger should a booster malfunction at launch.

libration. A real or apparent slow oscillation of a satellite in its orbit; in the case of the Moon, it allows us to see more than one hemisphere.

limb. The "edge" of the Moon as seen in the sky.

Line of Apsides. For the Moon, the line connecting the perigee and apogee points of its orbit.

Line of Nodes. The line connecting the nodes of an orbit.

lunar Farside. The side of the Moon that is not visible from Earth.

lunar Nearside. The side of the Moon that we can see from Earth.

lunar theory. An attempt to express all the motions of the Moon in one mathematical equation.

lunation. The length of time from the first appearance of the crescent Moon after "New" until the next such first appearance.

luni-solar calendar. A calendar that uses the apparent motions of both the Moon and Sun to keep track of the passage of time.

magma. Molten rock, usually at high temperatures and rich in silicates.

magnetic field. A region in which magnetic forces are present.

magnification. The number of times larger an image appears (as through a telescope) than the object appears to the naked eye.

mantle. That region of a planet's interior above the core and below the crust.

maria. Large, sparsely cratered, relatively smooth areas on the Moon.
mascons. Subsurface concentrations of unusually high density.
mass driver. A hypothetical electromagnetic "slingshot" for accelerating objects into orbit from the Moon without having to use rockets.
mass wasting. The movement of rock material downhill under the influence of gravity.
meteoroids. Small pieces of rock in space. A meteoroid that is seen as a flash of light as it enters the Earth's atmosphere is called a "meteor," and one that survives to be collected on the ground is called a "meteorite."
Metonic cycle. The period of 235 lunations, which is commensurate with a span of 19 years.
micrometeoroids. Tiny particles of rock or dust in space.
micrometer. An instrument for measuring small distances.
momentum. The product of an object's mass and velocity; in Newtonian mechanics, this quantity must be conserved.
moonbase. A proposed inhabitation on the Moon.
neap tide. The tide of lowest magnitude, occurring when the Sun and Moon are on opposite sides of the Earth and partially cancel each other's effects.
New Moon. That point at the end of a lunation at which the Moon is not visible because it is hidden by the glare of the Sun.
Newtonian telescope. A type of reflecting telescope in which the light from the secondary mirror is deflected through a hole in the side of the instrument.
Newton's Laws of Motion. Three laws formulated by Newton that describe mechanics and gravitation:

1. A body will tend to maintain its motion unless acted on by an external force.
2. The change in a body's momentum is equal to the force exerted on that body.
3. For every force (or action) there exists an equal and opposite force (or reaction).

nodes. The points in the Moon's orbit at which the orbit plane intersects the ecliptic.
nodical month. The same as the draconic month; the length of the month measured relative to the line of nodes (27.21 days).
objective. The principal lens or mirror in a telescope.
opposition. The point at which the Moon (or a planet) is on the opposite side of the Earth from the Sun.
orbital velocity. The velocity of a body in orbit about another.
paleomagnetic record. The record of fossil magnetism preserved in rocks.
parallax. An apparent change in the position of an object caused by a change in the position of the observer.
partial eclipse. An eclipse of the Sun or Moon during which that body is not totally obscured.

perigee. The point in the orbit of an Earth satellite at which it is closest to the Earth.

photomultiplier. A photosensitive cell that generates and amplifies an electrical signal when light falls on it.

planetary science. A relatively new distinct discipline applying a combination of astronomy, geology, and chemistry to the study of planets and planetesimals.

planetesimals. Any of the small solid bodies that have orbited the Sun at any time in the history of the solar system.

polarization. The degree to which the vibration planes of the light rays in a beam are aligned.

primum mobile. The "prime mover" or outermost sphere in the universe as conceived by Aristotle.

quadrature. The point at which the Earth–Sun line and the Earth–Moon line form a right angle.

radiogenic. Originating in radioactive decay.

radiogenic heating. Heat released in radioactive decay.

rare earth. Elements with atomic numbers 57 to 71.

rays. Bright linear features caused by material ejected at high speeds during crater formation.

reacting spheres. Invisible spheres that Aristotle hypothesized connected the spheres on which the planets rode.

reflecting telescope. A telescope in which mirrors are the optical components that form and direct the image.

refracting telescope. A telescope in which the principal optical components are lenses.

refraction. The bending of light rays as they pass from one medium to another, each medium having a different refractive index.

refractory elements. Elements having relatively high melting points; they would be the first to solidify out of a cooling nebula.

regolith. Loose fragments of rock and soil overlying solid bedrock.

rendezvousing. Manuevering two spacecraft near each other while in orbit.

resolution. Used in the text in the sense of how small a feature might be and still be seen or "resolved" in photographs.

retardation. The time lag between the time of the Moon's rising one night and its rising the next.

retroreflectors. Reflectors left on the Moon by the astronauts and used with lasers for measuring precisely the distance to the Moon.

retrorockets. Small rockets fired against the direction of motion of a spacecraft to cause its descent to a landing.

rilles. Any of several types of long, narrow depressions on the Moon's surface.

Roche limit. A distance from a primary body (e.g., the Earth) inside of which the forces holding a secondary body together (e.g., the Moon) will be overcome by the tidal force of the primary, at which point the secondary will be pulled apart.

rockets. Devices propelled by the ejection of the combustion products of solid or liquid fuel.

rubidium–strontium dating. Using the relative proportions of Rb–87 to Sr–87 (which is formed by the radioactive decay of Rb–87) to date rock.

saros cycle. The period of time, equal to 223 synodic months, after which time the synodic, nodical, and anomalistic months are nearly commensurate, thus allowing a particular pattern of eclipses to repeat itself.

scarps. High cliffs on the Moon.

secondary. In a reflecting telescope, a smaller mirror that redirects the image formed by the primary.

secondary craters. Craters formed by the ejecta from a primary impact.

secular acceleration. Slow, regular changes in the Moon's motions.

selenography. The mapping of the Moon.

shock metamorphosis. The alteration of rock caused by the violent impact of a meteoroid.

sidereal month. The orbital period of the Moon as measured against the background stars.

siderophilic elements. Elements with an affinity for the metallic phase.

sinii. Offshoots of the larger maria (from the Latin for "bay").

sinuous rilles. Meandering crevasses on the Moon (Also called serpentine rilles).

slump terraces. Terraces formed when part of a hillside gives way and slides downhill.

solar flare. An extremely energetic emission of radiation and charged particles from the Sun.

solar panels. Panels of photoelectric chips that produce electricity when light shines on them.

solar wind. A flux of charged particles from the Sun.

spacecraft. Vehicles designed to travel into space.

space probe. An unmanned spacecraft carrying instruments to photograph or otherwise obtain data on other bodies in the solar system or on interplanetary space.

space station. An inhabitation in space (likely in Earth orbit).

spacesuit. A suit specially designed to allow astronauts to work outside the confines of their spacecraft.

spectroscopy. Studying the spectra of objects or gases to determine composition, polarization, temperature, velocity, and other such properties.

spring tide. The tide of maximum amplitude that occurs when the Sun and Moon are on the same side of the Earth and their effects add.

stratigraphy. The study of different layers (or strata) in a rock formation, principally to assign relative ages.

sublunar sphere. In Aristotle's heliocentric conception of the universe, the only one of the "crystalline spheres" that traveled around the Earth.

suborbital. Not quite in orbit.

subsatellite. Small satellites put into orbit around the Moon by a departing Apollo spacecraft.

superconductivity. A property of certain materials that become very highly conducting at temperatures near absolute zero (−273° C.).

supernova. A tremendous explosion of a massive star in which the star becomes, for a brief period, many hundreds of thousands of times brighter.

synodic month. The month when measured by the Moon's position relative to the Sun (i.e., by its "phase").

syzygy. The configuration of the Earth, Moon, and Sun when the Moon is either new or full (i.e., when all three bodies are lined up in a straight line).

telescope. An instrument with lenses and/or mirrors that gathers more light than the naked eye and thus allows one to see very faint objects and more detail.

terminator. The line of shadow across the disk of the Moon (when it is viewed at any other phase than full).

terraforming. Ecologically modifying a satellite or planet to make it more Earth-like.

Third Quarter. That phase of the Moon three-quarters of the way through the lunation when we see the east half of the Moon's disk illuminated. (Also called "last quarter.")

thruster. A small rocket used for steering a spacecraft.

tidal bulge. One of the bulges of high tide on opposite sides of the Earth that are caused by the differential pull of the Moon.

tidally locked. Describing a satellite that has an orbit period equal to its rotation period.

total eclipse. An eclipse during which the entire disk of the Sun or Moon is covered.

Transient Lunar Phenomenon (TLP). A temporary change in the appearance of an area on the Moon.

transits. The passages of a celestial body across the meridian.

Tychonic System. Tycho's description of motion in the solar system in which the planets orbited the Sun, but the Sun and Moon still orbited the Earth.

Van Allen Radiation Belts. A region around the Earth where charged particles from the solar wind are trapped by the Earth's magnetic field.

variation. An irregularity in the motion of the Moon that is least at points of quadrature and syzygy.

vernal equinox. The point in the sky at which the Sun's path crosses the plane of the Earth's equator going from south to north.

vesicles. Bubble-shaped, smooth cavities formed by the expansion of gas in a magma.

volatile. An element that readily becomes gaseous.

vugs. Irregularly shaped cavities formed by the expansion of gas in a magma.

walled plains. Large craters on the Moon with diameters of about 100 miles or greater.

Waning Gibbous. That part of the lunation during which the lighted disk of the Moon is changing from Full to Third Quarter. (From Third Quarter to New we see a waning "crescent".)

Waxing Gibbous. That part of the lunation during which the lighted disk of the Moon is changing from First Quarter to Full. (From New to First Quarter we see a waxing "crescent".)

wrinkle ridge. Ridges that appear to be formed by overlapping edges of successive lava flows; also called mare ridges.

zodiac. Twelve primary constellations (or "houses") that map the yearly path of the Sun in the sky.

Note: Glossary compiled by Andrea Dobson-Hockey, New Mexico State University.

BIBLIOGRAPHY

The following books are in print. Most are moderately priced, and many are available from public libraries.

CHAPTER 2

E. C. Krupp, *Echoes of the Ancient Skies,* New York: Harper & Row, 1983.

CHAPTER 3

William K. Hartmann, *Moons and Planets,* 2nd ed., Belmont, Calif.: Wadsworth Publishing, 1983.

J. B. Zircher, *Total Eclipses of the Sun,* New York: Van Nostrand Reinhold, 1984.

J. L. E. Dreyer, *A History of Astronomy from Thales to Kepler,* New York: Dover, 1953.

CHAPTER 4

Galileo Galilei, *Discoveries and Opinions of Galileo,* New York: Doubleday, 1957.

P. Clay Sherrod and Thomas L. Koed, *A Complete Manual of Amateur Astronomy: Tools and Techniques for Astronomical Observations,* Englewood Cliffs, N.J.: Prentice-Hall, 1981.

Ernest H. Cherrington, *Exploring the Moon through Binoculars and Small Telescopes,* New York: Dover, 1983.

CHAPTER 5

Arthur Berry, *A Short History of Astronomy: From Earliest Times through the Nineteenth Century,* New York: Dover, 1961.

Agnes M. Clerke, *A Popular History of Astronomy in the Nineteenth Century,* St. Clair Shores, Minn.: Scholarly Press, 1908.

CHAPTER 6

Wernher Von Braun et al., *Space Travel: A History,* rev. ed., New York: Harper & Row, 1985.

Nicolas L. Johnson, *Handbook of Soviet Lunar and Planetary Exploration,* San Diego, Calif.: Univelt, 1979.

CHAPTER 7

Peter Cadogan, *The Moon—Our Sister Planet,* Cambridge, Eng.: Cambridge University Press, 1981.

The Earth's Moon, Washington, D.C.: National Geographic Society, 1969. Wall chart.

CHAPTER 8

Bruce Murray et al., *Earthlike Planets: Surfaces of Mercury, Venus, Earth, Moon, Mars,* New York: W. H. Freeman, 1981.

J. E. Guest and R. Greeley, *Geology on the Moon,* New York: Crane, Russak, 1977.

William B. Hubbard, *Planetary Interiors,* New York: Van Nostrand Reinhold, 1984. Technical.

Stuart R. Taylor, *Planetary Science: A Lunar Perspective,* Tucson, Ariz.: Lunar and Planetary Institute, 1982. Technical.

CHAPTER 9

Clark R. Chapman, *Planets of Rock and Ice: From Mercury to the Moons of Saturn,* New York: Charles Scribner's Sons, 1982.

Gerard O'Neill, *The High Frontier,* New York: Doubleday, 1982.

Astronomy, Milwaukee, Wis.: Astromedia. Monthly magazine.

Sky and Telescope, Cambridge, Mass.: Sky Publishing. Monthly magazine.

I might also mention that the Replogle Company of Chicago manufactures a handsome twelve-inch diameter *Lunar Globe.* One currently occupies a prominent place in my living room.

INDEX